Electronics

SECOND EDITION

A Self-Teaching Guide

HARRY KYBETT

A Wiley Press Book

JOHN WILEY & SONS, INC.
New York • Chichester • Brisbane • Toronto • Singapore

Publisher: Stephen Kippur
Editor: Theron Shreve
Managing Editor: Katherine Schowalter
Composition & Make-Up: G&H/Soho, Ltd.

Library of Congress Cataloging-in-Publication Data

Kybett, Harry.
   Electronics.

   (A Self-teaching guide)
   Includes index.
   1. Electronics. I. Title. II. Series: Wiley self-
teaching guides.
TK7816.K87 1986   621.381   85-26518
ISBN 0-471-00916-4

Printed in the United States of America

20 19 18 17

Preface

Modern electronics is a relatively new phenomenon. All of the things we see in the marketplace today that utilize electronics either did not exist before 1960 or were crude by today's standards. Some of the many examples of modern electronics in the home include the small but powerful pocket calculator, the personal computer, the miniature stereo radio/tape player, the VCR, automatic cameras, etc. Many industries have recently been founded and older industries revamped because of the availability and application of modern electronics in manufacturing processes as well as in the products themselves.

Modern electronics *is* the transistor and its offspring—the integrated circuit (IC) and the microprocessor. They have short-circuited much of traditional electronic theory, revolutionized its practice, and set the whole technology off on several new paths. This book is a first step to point you down those paths.

While many areas of our lives have become almost unbelievably complex, the study and practice of electronics in industry and as a hobby has surprisingly been made much simpler. *Electronics* takes advantage of this simplicity and covers only those areas actually needed in modern electronics rather than those found in the traditional textbooks. The traditional way of teaching electronics is often confusing; too many students are left feeling that the real core of electronics is mysterious and arcane, akin to black magic. This just is not so.

The arrangement and approach of this book is completely different from any other book on electronics. This book is for anyone who wants to get a start in electronics. It focuses on the very few important principles you need in working with electronics and it concentrates on the practical side of the subject, stressing its usefulness as a technology. A "question and answer" approach is used to lead the student into simple but pertinent experiments. Almost no math is used, and none beyond first year algebra level. In addition, the usual chapters on semiconductor physics have been completely omitted, since these are not needed in the early stages of electronics.

Electronics is a very easy technology, which anyone can understand with very little effort, once shown what is really going on. This book focuses on how to apply the few basic principles which are the basis of modern electronic practice.

How to Use This Book

This book assumes some knowledge of basic electricity. If you have read a textbook or taken a course on electricity, or if you have worked with electricity, you probably have the prerequisites. If not, you should read Charles Ryan's *Basic Electricity* (also a Self-Teaching Guide from John Wiley & Sons) or some other basic book on electricity to get the necessary background for this book. When you start to read *Electronics*, Chapters 1 and 5 will allow you to test and review the necessary basics of electricity as you read. You should read the chapters in order, because often later material depends on concepts and skills applied in earlier chapters.

Electronics is presented in a self-teaching format that allows you to learn easily and at your own pace. The material is presented in numbered sections called frames. Each frame presents some new information and gives you a question, problem, or experiment to try. To learn most effectively, you should try to answer each question fully on your own. Then compare your answer with the correct answer given following the line of dashes. If you miss a question, correct your answer and then go on; if you miss many in a row, go back and review the previous section. Otherwise you may miss the point of the new material. Also, try to do all the experiments. They are very easy and are all pertinent to the subject matter.

When you reach the end of a chapter, evaluate your learning by taking the Self-Test. If you miss any questions, review the indicated parts of the chapter again. If you do well on the Self-Test, you're ready to go on to the next chapter. You may also find the Self-Test useful as a review before you start the next chapter. At the end of the book is a Final Self-Test that will allow you to assess your overall learning.

Go through this book at your own pace and use it for only about 20 minutes at a time. You can work through this book alone or you can use it in conjunction with a course. If used alone, it serves as an introduction to electronics—it is not a complete course. At the end of the book are some comments on further reading and applications. Also at the back of the book is a table of symbols and abbreviations, which will be useful for reference and review.

Now you're ready to learn *Electronics!*

Acknowledgments

This edition would not have been possible without the able assistance provided by Professor Joseph M. Farren of the Electronic Engineering Technology Department of the University of Dayton. He revised the original edition and provided the new material and exercises necessary to update this edition.

Thanks are also due to the following reviewers: Earl C. Iselin, Jr.; Stan Antosz; and Jacques Trudeau.

Contents

CHAPTER ONE

DC Pre-Test and Review

Electronics cannot be studied without first studying the basics of electricity. This chapter is a review and pre-test on those aspects of DC (direct current) which apply to electronics. By no means does it cover the whole DC theory, but merely those topics which are essential to simple electronics. This pre-test will review:

- current flow;
- potential or voltage difference;
- Ohm's law;
- resistors in series and parallel;
- power;
- small currents;
- resistance graphs;
- Kirchhoff's voltage and current laws;
- voltage and current dividers;
- switches;
- capacitor charging and discharging;
- capacitors in series and parallel.

CURRENT FLOW

1. Electrical and electronic devices work because of an electric current. What is an electric current? _____
    ― ― ― ― ― ― ― ― ― ― ― ― ― ― ―

An electric current is a flow of electric charge. The electric charge usually consists of negatively charged electrons; however, with semiconductors there will also be positive charge carriers called "holes."

2. Can you write at least three ways an electron flow—or current—can be generated? _____

_____

– – – – – – – – – – – – – – – – – –

The main ways are chemically, magnetically, thermally, piezoelectrically, and photoelectrically. Other ways are possible. In practice the most common are: (1) chemically—by using a battery, and (2) magnetically—by using a generator.

3. Most of the simple examples in this book will contain a battery as the voltage source. As such, the source provides a potential difference to a circuit that will enable a current to flow. An electric current is a flow of electric charge. In the case of a battery, electrons are the electric charge and they will flow from the terminal that has an excess number of electrons to the terminal that has a deficiency of electrons. This flow takes place in the circuit that is externally connected to the battery terminals. It is this difference of charge that creates the potential difference in the battery; the electrons are trying to balance the difference.

   Since electrons have a negative charge, they actually flow from the negative terminal and return to the positive terminal. This direction of flow is called *electron flow*. Most books, however, use current flow, which is in the opposite direction. It is referred to as *conventional current flow* or simply *current flow*. In this book, we will observe conventional current flow in all our circuits.

   Later in the text you will see that many semiconductor devices have a symbol that contains an arrowhead pointing in the direction of conventional current flow.

   (a) Draw arrows to show the current flow in this figure. Note the symbol for the battery and its polarity.

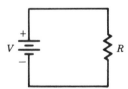

   (b) What indicates that a potential difference is present? _____

_____

(c)   What does the potential difference cause? _____

_____

(d)   What will happen if the battery is reversed? _____

_____

- - - - - - - - - - - - - - - -

(a)

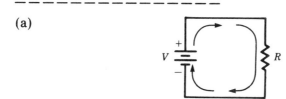

(b)   The battery symbol indicates that a difference of potential is being
       supplied to the circuit.
(c)   It causes current to flow if there is a complete circuit present as
       shown in the figure.
(d)   The current will flow in the opposite direction.

OHM'S LAW

4.   Ohm's law states the fundamental relationship between voltage, current,

and resistance. Write its algebraic formula. _____
- - - - - - - - - - - - - - - -

$V = R \times I$

This is the most basic equation in electricity, and you should know it
well. Note that some textbooks state Ohm's Law as $E = IR$. E and V are
both symbols for voltage. E is generally used for sources and V is used
for voltage drops or differences in potential as well as for sources. In this
book we will use V throughout. Also, in this formula, resistance is the
opposition to current flow. Note that when resistance is large, the current
will be small.

5.   The exercises in this frame depend on Ohm's law. In each, R and I are
      given; you must find V.

(a) $R = 20$ ohms $I = 0.5$ amperes $V =$ _____

(b) $R = 560$ ohms $I = 0.02$ amperes $V =$ _____

(c) $R = 1000$ ohms $I = 0.01$ amperes $V =$ _____

(d) $R = 20$ ohms $I = 1.5$ amperes $V =$ _____
- - - - - - - - - - - - - - - -

    (a)   10 volts
    (b)   11.2 volts
    (c)   10 volts
    (d)   30 volts

6.   In these exercises, V and R are given. Find I.

    (a) $V = 1$ volt $R = 2$ ohms $I = $ _____

    (b) $V = 2$ volts $R = 10$ ohms $I = $ _____

    (c) $V = 10$ volts $R = 3$ ohms $I = $ _____

    (d) $V = 120$ volts $R = 100$ ohms $I = $ _____

    – – – – – – – – – – – – – – –

    (a)   0.5 amperes
    (b)   0.2 amperes
    (c)   3.3 amperes
    (d)   1.2 amperes

7.   In these exercises, V and I are given. Find R.

    (a) $V = 1$ volt $I = 1$ ampere $R = $ _____

    (b) $V = 2$ volts $I = 0.5$ ampere $R = $ _____

    (c) $V = 10$ volts $I = 3$ amperes $R = $ _____

    (d) $V = 50$ volts $I = 20$ amperes $R = $ _____

    – – – – – – – – – – – – – – –

    (a)   1 ohm
    (b)   4 ohms
    (c)   3.3 ohms
    (d)   2.5 ohms

8.   Work through these examples. In each case two factors are given and you must find the third.

    (a)   12 volts and 10 ohms. Find the current. _____

    (b)   24 volts and 8 amperes. Find the resistance. _____

    (c)   5 amperes and 75 ohms. Find the voltage. _____

    – – – – – – – – – – – – – – –

    (a)   1.2 amperes
    (b)   3 ohms
    (c)   375 volts

RESISTORS IN SERIES

9. Resistors can be connected in series. The figure below shows two resistors in series.

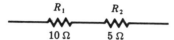

R_1 R_2

$10 \, \Omega$ $5 \, \Omega$

What is their total resistance? _____

$R_T = R_1 + R_2 = 10$ ohms $+ 5$ ohms $= 15$ ohms

The total resistance is often called the "equivalent series resistance"—R_{eq}.

RESISTORS IN PARALLEL

10. Resistors can be connected in parallel, as in the figure below.

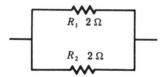

R_1 2Ω

R_2 2Ω

What is the total resistance here? _____

$\frac{1}{R_T} = \frac{1}{R_1} + \frac{1}{R_2} = \frac{1}{2} + \frac{1}{2} = \frac{1}{1} = 1$ ohm

This is often called the "equivalent parallel resistance." The formula is very easy to use with a calculator.

11. The simple formula from frame 10 can be extended to include as many resistors as desired. What is the formula for three resistors in parallel?

_____

$\frac{1}{R_T} = \frac{1}{R_1} + \frac{1}{R_2} + \frac{1}{R_3}$

12. In the following exercises the resistors are connected in parallel. Find their total or equivalent resistance.

(a) $R_1 = 1$ ohm $R_2 = 1$ ohm $R_T =$ _____

(b) $R_1 = 1000$ ohms $R_2 = 500$ ohms $R_T =$ _____

(c) $R_1 = 3600$ ohms $R_2 = 1800$ ohms $R_T =$ _____

(a)   0.5 ohms
(b)   333 ohms
(c)   1200 ohms

Note that R_T is always smaller than the smallest of the resistors in parallel.

POWER

13.   When current flows through a resistor it dissipates power, usually in the form of heat. Power is expressed in terms of watts. What is the formula for power? _____

— — — — — — — — — — — — —

$P=VI$ or $P = I^2R$ or $P = \dfrac{V^2}{R}$

14.   In these exercises, find the power dissipated by the resistor when the voltage and currents given are applied to the resistor.

(a) $V = 10$ volts $I = 3$ amperes $P =$ _____

(b) $V = 100$ volts $I = 5$ amperes $P =$ _____

(c) $V = 120$ volts $I = 10$ amperes $P =$ _____

— — — — — — — — — — — — —

(a)   30 watts
(b)   500 watts, or 0.5 kilowatts
(c)   1200 watts, or 1.2 kilowatts

15.   Now find the power dissipated by the resistors for frames 5 and 6, parts a and b.

5(a) _____   5(b) _____

6(a) _____   6(b) _____

— — — — — — — — — — — — —

5(a) 5 watts   5(b) 0.224 watts
6(a) 0.5 watts   6(b) 0.4 watts

16.   Resistors used in electronics generally are manufactured in standard values with regard to resistance and power rating. A table of standard resistance values is given in the appendix. Quite often, when a certain resistance value is needed in a circuit, you must choose the closest standard value. You will see this in numerous examples in this book.

You must also choose a resistor with the power rating in mind. You should never place a resistor in a circuit that would require that resistor to dissipate more power than its rating.

If standard power ratings for composition resistors are 1/4, 1/2, 1, and 2 watts, what power ratings should be selected for the resistors that were used for the calculations in frame 15?

For 5 watts _____   For 0.224 watts_____

For 0.5 watts _____   For 0.4 watts_____

— — — — — — — — — — — — — —

For the 5 watt requirement, you would have to select a power resistor rather than a composition resistor. The other resistors would have to be 1/4, 1/2, and 1/2 watt resistors. In electronics, most applications would require low power resistors such as composition resistors. For higher power levels other types of resistors are available.

SMALL CURRENTS

17. Although currents much larger than 1 ampere are found in heavy industrial equipment, in most electronic work only fractions of an ampere are found.

    (a) What is the meaning of the term *milliampere?*_____

    _____

    (b) What does the term *microampere* mean? _____

    _____

    — — — — — — — — — — — — — —

    (a) A milliampere is one-thousandth of an ampere, that is, 1/1000 or 0.001 amperes. It is abbreviated mA.
    (b) A microampere is one-millionth of an ampere, that is, 1/1,000,000 or 0.000001 amperes. It is abbreviated μA.

18. In electronics the values of resistance normally encountered are quite high. Often thousands of ohms and occasionally even millions of ohms are used.

    (a) What does kΩ mean when it refers to a resistor? _____

    _____

    (b) What does MΩ mean when it refers to a resistor? _____

    _____

    — — — — — — — — — — — — — —

    (a)   Kilohm (k = kilo, Ω = ohm). The resistance value is measured in thousands of ohms. Thus, 1 kΩ = 1000.ohms, 2 kΩ = 2000 ohms, and 5.6 kΩ = 5600 ohms.

    (b)   Megohm (M = mega, Ω = ohm). The resistance is measured in millions of ohms. Thus, 1 MΩ = 1,000,000 ohms, and 2.2 MΩ = 2,200,000 ohms.

19.   The following exercise is typical of many performed in transistor circuits. 6 V will be applied across a resistor and 5 mA of current is required to flow through the resistor.

        What value of resistance must be used and what power will it dissipate?

$R =$ _____ $P =$ _____

— — — — — — — — — — — — — — — — —

$$R = \frac{V}{I} = \frac{6\text{ V}}{5\text{ mA}} = \frac{6}{0.005} = 1200 \text{ ohms} = 1.2 \text{ k}\Omega$$

$$P = V \times I = 6 \times 0.005 = 0.030 \text{ watts} = 30 \text{ mW}$$

20.   Now try these two simple examples.

    (a)   50 volts and 10 mA. Find the resistance. _____

    (b)   1 volt and 1 MΩ. Find the current. _____

— — — — — — — — — — — — — — — — —

    (a)   5 kΩ
    (b)   1 μA

THE GRAPH OF RESISTANCE

21.   The voltage drop across a resistance and the current flowing through it can be plotted on a simple graph. This graph is called a *V-I* curve.

        Consider a simple circuit in which a battery is connected across a 1 kΩ resistor.

    (a)   Find the current flowing if the battery is 10 V. _____

    (b)   Now use a 1 V battery and find the current. _____

    (c)   Now find the current when a 20 V battery is used. _____

— — — — — — — — — — — — — — — —

    (a)   10 mA
    (b)   1 mA
    (c)   20 mA

22. Plot the points of battery voltage and current flow from frame 21 on the graph below, and connect them together.

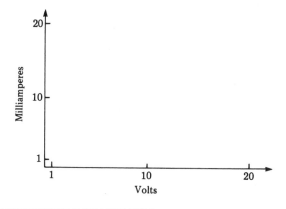

You should have drawn a straight line, as in the graph below.

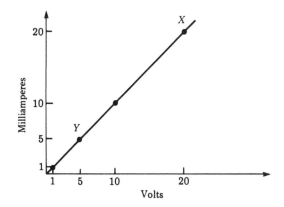

Sometimes we want to calculate the *slope* of the line on a graph. To do this, pick two points; call them X and Y.

Let $X = 20$ V and 20 mA
Let $Y = 5$ V and 5 mA

Thus the slope $= \dfrac{20\ \text{V} - 5\ \text{V}}{20\ \text{mA} - 5\ \text{mA}} = \dfrac{15\ \text{V}}{15\ \text{mA}} = 1\ \text{k}\Omega$

In other words, the slope of the line is equal to the resistance.

Later you will learn about *V-I* curves for other components. They have several uses, and often they are not straight lines.

THE VOLTAGE DIVIDER

23. The circuit shown below is called a *voltage divider*. It is the basis for many important theoretical and practical ideas throughout the entire field of electronics.

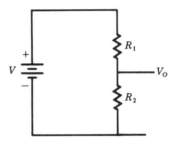

The important point in this circuit is V_o—the *voltage drop* across R_2.

What is the formula for this voltage drop? _____

– – – – – – – – – – – – – – – – – –

$$V_o = V \times \frac{R_2}{R_1 + R_2}$$

Note that $R_1 + R_2 = R_T$, the actual total resistance of the circuit.

24. A simple example will demonstrate the use of this formula. In the circuit below, find the voltage drop (V_O) across R_2.

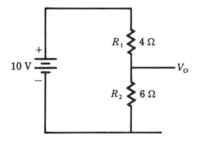

– – – – – – – – – – – – – – – – – –

$$V_O = V \times \frac{R_2}{R_1 + R_2}$$

$$= 10 \times \frac{6}{4 + 6}$$

$$= 10 \times \frac{6}{10}$$

$$= 6 \text{ volts}$$

25. Now try these problems. Find V_O in each case.

  (a) $V = 1$ volt $R_1 = 1$ ohm $R_2 = 1$ ohm

  (b) $V = 6$ volts $R_1 = 4$ ohms $R_2 = 2$ ohms

  (c) $V = 10$ volts $R_1 = 3.3$ kΩ $R_2 = 5.6$ kΩ

  (d) $V = 28$ volts $R_1 = 22$ kΩ $R_2 = 6.2$ kΩ

  - - - - - - - - - - - - - - - -

  (a) 0.5 volts
  (b) 2 volts
  (c) 6.3 volts
  (d) 6.16 volts

The output voltage from the voltage divider is always less than the applied voltage. It is said to be "attenuated," and the voltage divider is a very simple "attenuator." Attenuators are very important in communications circuits.

26. Again, using the voltage divider equation, find the voltage across the 22k resistor for part d of frame 25. What do you get if you add this

  value of voltage to the voltage across the 6.2k resistor? _____

  - - - - - - - - - - - - - - - -

  21.84 volts
  The sum is 28 volts.

  Note that the voltages across the two resistors add up to the supply voltage. This is an example of Kirchhoff's voltage law (KVL), which simply means that the voltage supplied to a circuit must equal the sum of the voltage drops in the circuit. It can be stated in a more complicated fashion, but that is left to other texts. In this book, KVL will often be used without actual reference to the law.
    Also notice that the most voltage goes to the largest resistor in a series string of resistors.

THE CURRENT DIVIDER

27. In the circuit shown below the current splits or divides between the two resistors which are placed in parallel.

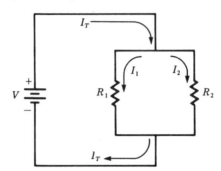

I_T splits into the individual currents I_1 and I_2 and then these recombine to form I_T. Indicate which of the relationships below are valid for this circuit.

___ (a) $V = R_1 I_1$

___ (b) $V = R_2 I_2$

___ (c) $R_1 I_1 = R_2 I_2$

___ (d) $I_1/I_2 = R_2/R_1$

— — — — — — — — — — — — — — — —

All of them are valid. The last one, d, shows that the current divides in the "inverse ratio of the resistance values."

28. When solving current divider problems, the following steps should be used.

(1) Set up the ratio of the resistors and currents.

$$I_1/I_2 = R_2/R_1$$

(2) Rearrange the ratio to give I_1 in terms of I_2.

$$I_1 = I_2 \times \frac{R_2}{R_1}$$

(3) From the fact that $I_T = I_1 + I_2$, express I_T in terms of I_1 only.

(4) Now find I_1 or I_2.

(5) Now find the remaining current (I_2 or I_1).

In the following example two resistors in parallel and the total current flowing are given. The object is to find the currents through each individual resistor.

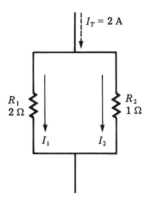

Working through the steps as shown:

(1) $I_1/I_2 = R_2/R_1 = 1/2$

(2) $I_2 = 2I_1$

(3) $I_T = I_1 + I_2 = I_1 + 2I_1 = 3I_1$

(4) $I_1 = I_T/3 = 2/3$ A

(5) $I_2 = 2I_1 = 4/3$ A

Now try these problems. In each case the total current and the two resistors are given. Find I_1 and I_2.

(a) $I_T = 30$ mA, $R_1 = 12$ kΩ, $R_2 = 6$ kΩ

_____

(b) $I_T = 133$ mA, $R_1 = 1$ kΩ, $R_2 = 3$ kΩ

_____

(c) What do you get if you add I_1 and I_2?

_____

— — — — — — — — — — — — — — — —

(a) $I_1 = 10$ mA, $I_2 = 20$ mA
(b) $I_1 = 100$ mA, $I_2 = 33$ mA
(c) They add back together to give you the total current supplied to the parallel circuit.

Note that question (c) is actually a demonstration of Kirchhoff's current law (KCL). This law simply stated says that the total current entering a junction in a circuit must equal the sum of the currents leaving that junction. This law will also be used on numerous occasions in later chapters. KVL and KCL together form the basis for many techniques and methods of analysis that are used in the application of circuit analysis.

Note also that the most current goes to the smallest resistor in a parallel circuit. Check your results above to verify this.

29. Another equation that is helpful in dividing current in a two branch parallel circuit is as follows:

$$I_1 = \frac{(I_T)(R_2)}{(R_1 + R_2)}$$

Now you write the equation for the current I_2.

_____

Check the answers for the previous frame using these equations.

$$I_2 = \frac{(I_T)(R_1)}{(R_1 + R_2)}$$

We see that the current through one branch of a two branch circuit is equal to the total current times the resistance of the opposite branch divided by the sum of the resistances of both branches. This is an easy formula to remember. We can also see from the formula that the most current goes to the smallest resistance. Checking the answers from the previous frame should point this out.

SWITCHES

30. A mechanical switch is a device which completes or breaks a circuit. The most familiar use is that of applying power to turn a device on or off. A switch can also permit a signal to pass from one place to another, prevent its passage, or route a signal to one of several places.

   Here we will review two types of switches. The first is the simple on-off switch, also called a *single pole single throw* switch. The second is the *single pole double throw* switch. The circuit symbols for each are shown here.

ON-OFF switch

in the OFF position

Single pole double throw or SPDT switch

   Two important facts about switches must be known. (1) A CLOSED (or ON) switch has the total circuit current flowing through it; there is *no* voltage drop across its terminals. (2) An OPEN (or OFF) switch has *no* current flowing through it; the full circuit voltage appears across its terminals. Look at the circuit below.

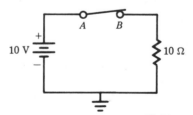

(a) What is the current flowing through the switch? _____

(b) What is the voltage at point A and point B with respect to ground? _____

(c) What is the voltage drop across the switch? _____

- - - - - - - - - - - - - - - - - -

(a) $\dfrac{10 \text{ V}}{10 \text{ ohms}} = 1$ ampere

(b) $V_A = V_B = 10$ V. It takes two points to define a voltage.

(c) 0 V. There is no voltage drop as both terminals are at the same voltage.

31. Now look at this circuit.

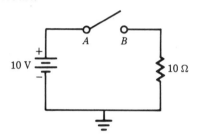

(a) What is the voltage at point A and point B? _____

(b) How much current is flowing through the switch? _____

(c) What is the voltage drop across the switch? _____

- - - - - - - - - - - - - - - - - -

(a) $V_A = 10$ V; $V_B = 0$ V

(b) No current is flowing since the switch is open.

(c) 10 V. If the switch is open it actually represents an infinite resistance, and since the most voltage goes to the largest resistance, it will all go to the infinite resistance or open circuit.

32. In the simple circuit shown below, illustrating a single pole double throw switch, either lamp A or lamp B will be lit, depending on the position of the switch.

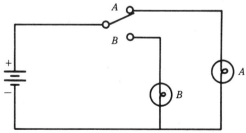

(a)   In the position shown, which lamp is lit? _____

(b)   Can both lamps be lit simultaneously? _____

_ _ _ _ _ _ _ _ _ _ _ _ _ _ _

(a)   lamp A

(b)   no, one or the other must be off

CAPACITORS IN A DC CIRCUIT

33.   Capacitors are used extensively in electronics. Although their main use is with AC signals there are certain specific areas of DC where they are very important. Their main DC use in electronics is to become charged and hold the charge; hence the following basic facts are essential.

In the figure below, the capacitor will charge when the switch is closed.

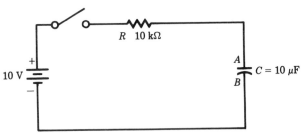

What will be the final voltage to which it will charge? _____

_ _ _ _ _ _ _ _ _ _ _ _ _ _ _ _

It will charge up to 10 V. It will actually charge up to the voltage that would appear across an open circuit located at the same place where the capacitor is located.

34.   How long does it take to reach this voltage? This is a most important question, with many practical applications. To find the answer the *time constant* of the circuit must be known.

(a)   What is the formula for the time constant of this type of circuit?

_____

(b)   What is the time constant for the circuit shown in frame 30?

_____

(c)   How long does it take the capacitor to reach 10 V?

_____

(d)   To what voltage level does it charge in one time constant?

_____

_ _ _ _ _ _ _ _ _ _ _ _ _ _ _

(a) $T = RC$

(b) $T = 10 \text{ k}\Omega \times 10 \text{ }\mu\text{F} = 0.1$ seconds

(c)   approximately 5 time constants, or about 0.5 seconds

(d)   63% of the final voltage, or about 6.3 V

35. Before the switch is closed, the capacitor will be uncharged. When a capacitor is uncharged or discharged, it has the same voltage on both plates.

(a)   What will be the voltage on plate A and plate B in the circuit diagrammed in frame 30 before the switch is closed? _____

(b)   When the switch is closed, what will happen to the voltage on plate A? _____

(c)   What will happen to the voltage on plate B? _____

(d)   What will be the voltage on plate A after one time constant?

_____

– – – – – – – – – – – – – – – – –

(a)   both will be at 0 V if the capacitor is totally discharged

(b)   it will rise towards 10 V

(c)   it will stay at 0 V

(d)   about 6.3 V

36. Capacitor charges can be drawn on a graph of the voltage rise plotted against time. *A capacitor charging graph* is shown here with several points marked.

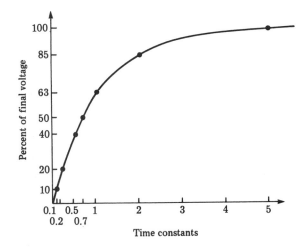

(a) What type of curve is this called? _____

(b) What is it used for? _____

- - - - - - - - - - - - - - - -

(a) It is called an *exponential* curve.
(b) It is used to calculate how far a capacitor has charged in a given time.

Practical applications of capacitor charging include timing circuit, preventing distortion in audio and communications circuit, and in deliberately altering and generating electronic signals.

37. In the examples below a resistor and a capacitor are in series. Calculate the time constant (T), how long it takes the capacitor to fully charge, and the voltage level after one time constant if a 10 V battery is used.

(a) $R = 1 \ k\Omega$ $C = 1000 \ \mu F$

_____

(b) $R = 330 \ k\Omega$ $C = 0.05 \ \mu F$

_____

- - - - - - - - - - - - - - - -

(a) $T = 1$ second; charge time = 5 seconds; $V_C = 6.3$ V
(b) $T = 16.5$ ms; charge time = 82.5 ms; $V_C = 6.3$ V (The abbreviation ms indicates milliseconds.)

38. Now look at this circuit.

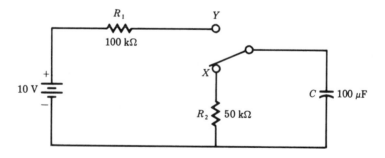

(a) With the switch in position X, what is the voltage on each plate of the capacitor? _____

(b) The switch is moved to position Y. The capacitor begins to charge. What is its charging time constant? _____

(c) How long does it take to fully charge? _____

- - - - - - - - - - - - - - - -

(a)   0 V

(b) $T = RC = (100 \text{ k}\Omega)(100 \ \mu F) = 10$ secs

(c)   approximately 50 seconds

39. When the capacitor is fully charged, the switch is put back to position X.

(a)   What is its discharge time constant? _____

(b)   How long does it take to fully discharge? _____

(c)   During the period of time that it might take to make the switch

from Y back to X, what is the capacitor doing? _____

_____

− − − − − − − − − − − − − − − −

(a) $T = RC = (50 \text{ k}\Omega)(100 \ \mu F) = 5$ seconds (discharging through the
50 kΩ resistor)

(b)   approximately 25 seconds

(c)   The capacitor holds or stores the 10 volts. It has the capacity to
store charge. We say from this that it has the ability to store
energy, since the charge shows up in an external circuit as a
potential difference. This stored energy can then be used in the
circuit at a later time.

40. Capacitors can be connected in parallel, as in the two figures below.

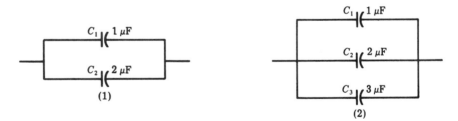

(1)   (2)

(a)   What is the formula for the total capacitance? _____

(b)   What is the total capacitance in figure 1? _____

(c)   What is the total capacitance in figure 2? _____

− − − − − − − − − − − − − − − −

(a) $C_T = C_1 + C_2 + C_3 + \ldots + C_N$

(b) $C_T = 1 + 2 = 3 \ \mu F$

(c) $C_T = 1 + 2 + 3 = 6 \ \mu F$

In other words, the total capacitance is found by simple addition of the
capacitor values.

41.   Capacitors can be placed in series, as in the figure below.

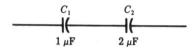

C_1 C_2

1 μF   2 μF

(a)   What is the formula for the total capacitance? _____

(b)   In the above figure, what is the total capacitance? _____

- - - - - - - - - - - - - -

(a) $1/C_T = 1/C_1 + 1/C_2 + 1/C_3 + \ldots + 1/C_N$
(b) $1/C_T = 1/1 + 1/2 = 3/2$; thus, $C_T = 2/3$ μF

42.   In each of these examples the capacitors are placed in series. Find the total capacitance.

(a) $C_1 = 10$ μF $C_2 = 5$ μF   _____

(b) $C_1 = 220$ μF $C_2 = 330$ μF $C_3 = 470$ μF   _____

(c) $C_1 = 0.33$ μF $C_2 = 0.47$ μF $C_3 = 0.68$ μF   _____

- - - - - - - - - - - - - -

(a)   3.3 μF
(b)   103.06μF
(c)   0.15 μF

SUMMARY

The few simple principles reviewed in this chapter are those needed to begin the study of electronics. They are listed here in summary form. Other important principles and ideas will be introduced throughout the book.

- The basic electrical circuit consists of a source (voltage), a load (resistance), and a path (conductor or wire).

- The voltage represents a charge difference.

- If the circuit is a complete circuit, then electrons will flow in what we call current flow. The resistance offers opposition to current flow.

- The relationship of V, I, and R is given by Ohm's Law.

$$I = \frac{V}{R}$$

- Resistance could be a combination of resistors in series, in which case we add them together to get the total resistance.

$$R_T = R_1 + R_2 + \ldots + R_N$$

- Resistance could be a combination of resistors in parallel, in which case we find the total from the following formula.

$$\frac{1}{R_T} = \frac{1}{R_1} + \frac{1}{R_2} + \ldots + \frac{1}{R_N}$$

- The power delivered by a source is found by:

$$P = VI$$

- The power dissipated by a resistance is:

$$P = I^2R = \frac{V^2}{R} = VI$$

- If you know the total applied voltage, E, the voltage across one resistor in a series string of resistors is found from the voltage divider formula.

$$V_1 = \frac{ER_1}{R_T}$$

- The current through one resistor in a two resistor parallel circuit with the total current known is found from the current divider formula.

$$I_1 = \frac{I_T R_2}{(R_1 + R_2)}$$

- Kirchhoff's voltage law (KVL) relates the voltage drops in a series circuit to the total applied voltage.

$$E \text{ or } V_T = V_1 + V_2 + \ldots + V_N$$

- Kirchhoff's current law (KCL) relates the currents at a junction in a circuit by saying that the sum of the input currents equals the sum of the output currents. For a simple parallel circuit, this becomes

$$I_T = I_1 + I_2 + \ldots + I_N$$

where I_T is the input current.

- A switch in a circuit becomes the control device that directs the flow of current or in many cases allows it to flow.

- Capacitors are used to store electric charge in a circuit. They also allow current or voltage to change at a controlled pace. The circuit time constant is found from the formula

$$T = RC$$

- At one time constant in an RC circuit the values for current and voltage have reached 63% of their final values. At five time constants they have reached their final values.

- Capacitors in parallel are added to get a total value.

$$C_T = C_1 + C_2 + \ldots + C_N$$

- Capacitors in series are treated the same as resistors in parallel to find a total capacitance.

$$\frac{1}{C_T} = \frac{1}{C_1} + \frac{1}{C_2} + \ldots + \frac{1}{C_N}$$

SELF-TEST

The problems and questions below will test your understanding of the basic principles presented in this chapter. You will need a separate sheet of paper for your calculations. Compare your answers with the answers provided following the test. You will find that many of the problems can be worked more than one way.

Questions 1–5 use the circuit shown below. Find the unknown values indicated using the values given.

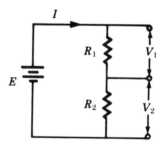

1. $R_1 = 12$ ohms, $R_2 = 36$ ohms, $E = 24$ V

 $R_T =$ _____, $I =$ _____

2. $R_1 = 1$ kΩ, $R_2 = 3$ kΩ, $I = 5$ mA

 $V_1 =$ _____, $V_2 =$ _____, $E =$ _____

3. $R_1 = 12$ kΩ, $R_2 = 8$ kΩ, $E = 24$ V

 $V_1 =$ _____, $V_2 =$ _____

4. $E = 36$ V, $I = 250$ mA, $V_1 = 6$ V

 $R_2 =$ _____

5. Now go back to problem 1 and find the power dissipated by each resistor and the total power delivered by the source.

 $P_1 =$ _____, $P_2 =$ _____, $P_T =$ _____

Questions 6–8 will use the following circuit. Again, find the unknowns using the given values.

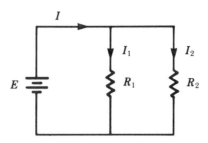

6. $R_1 = 6$ kΩ, $R_2 = 12$ kΩ, $E = 20$ V

    $R_T =$ _____, $I =$ _____

7. $I = 2$ A, $R_1 = 10$ ohms, $R_2 = 30$ ohms

    $I_1 =$ _____, $I_2 =$ _____

8. $E = 12$ V, $I = 300$ mA, $R_1 = 50$ ohms

    $R_2 =$ _____, $P_1 =$ _____

9. What is the maximum current that a 220 ohm resistor can safely have
    if its power rating is 1/4 watts?

    $I_{MAX} =$ _____

10. In a series RC circuit the resistance is 1 kΩ, the applied voltage is 3 V,
    and the time constant is to be 60 μsec.

    (a) What is the required value of C?

        $C =$ _____

    (b) What will be the voltage across the capacitor 60 μsec after the
        switch is closed?

        $V_C =$ _____

    (c) At what time will the capacitor be fully charged?

        $T =$ _____

11. In the circuit shown, when the switch is at position 1, the required time
    constant is 4.8 ms.

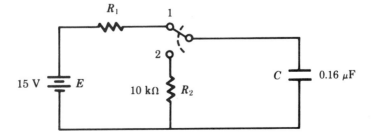

(a) What should be the value of resistor R_1?

$R_1 = $ _____

(b) What will be the voltage on the capacitor when it is fully charged, and how long will it take to reach this voltage?

$V_C = $ _____, $T = $ _____

(c) After the capacitor is fully charged, the switch is thrown to position 2. What is the discharge time constant and how long will it take to completely discharge the capacitor?

$TC = $ _____, $T = $ _____

12. Three capacitors are available with the following values:

$C_1 = 8 \ \mu F; \ C_2 = 4 \ \mu F; \ C_3 = 12 \ \mu F.$

(a) What is C_T if all three are connected in parallel?

$C_T = $ _____

(b) What is C_T if they are connected in series?

$C_T = $ _____

(c) What is C_T if C_1 is in series with the parallel combination of C_2 and C_3?

$C_T = $ _____

Answers to Self-Test

If your answers do not agree with those given below, review the frames indicated in parentheses before you go on to the next chapter. It is assumed that Ohm's Law is well known, so frame 4 will not be referenced.

1.	$R_T = 48$ ohms, $I = 0.5$ A	(frame 9)
2.	$V_1 = 5$ V, $V_2 = 15$ V, $E = 20$ V	(frame 26)
3.	$V_1 = 14.4$ V, $V_2 = 9.6$ V	(frame 23)
4.	$R_2 = 120$ ohms	(frames 9 and 23)
5.	$P_1 = 3$ W, $P_2 = 9$ W, $P_T = 12$ W	(frame 13)
6.	$R_T = 4 \ k\Omega, \ I = 5$ mA	(frame 10)
7.	$I_1 = 1.5$ A, $I_2 = 0.5$ A	(frame 29)
8.	$R_2 = 200$ ohms, $P_1 = 2.88$ W	(frames 10 and 13)
9.	$I_{MAX} = 33.7$ mA	(frames 15–16)

10. (a) $C = 0.06\ \mu F$
    (b) $V_C = 1.9$ V
    (c) $T = 300\ \mu sec$ (frames 33–37)

11. (a) $R_1 = 30$ kΩ
    (b) $V_C = 15$ V, $T = 24$ ms
    (c) $TC = 1.6$ ms, $T = 8.0$ ms (frames 38–39)

12. (a) $24\ \mu F$
    (b) $2.18\ \mu F$
    (c) $5.33\ \mu F$ (frames 41–42)

CHAPTER TWO

The Diode

The main characteristic of the diode is that it conducts electricity in one direction only. Historically the first vacuum tube was a diode; it was also known as a *rectifier*. The modern diode is a semiconductor device. It is used in all applications where the older vacuum tube diode was used, but it has the advantages of being much smaller, easier to use, and less expensive.

The term *semiconductor* describes a class of devices whose properties do not permit them to be classified with either conductors or insulators. Under the correct conditions they will conduct an electric current in a well defined and controlled manner; they are the basic material of all modern electronic circuits. We have to study the semiconductor diode as an introduction to the transistor.

When you complete this chapter you will be able to:

- specify the uses of diodes in DC circuits;

- determine from a diagram whether a diode is forward or reverse biased;

- recognize the characteristic *V-I* curve for a diode;

- specify the knee voltage for a silicon or a germanium diode;

- calculate current and power dissipation in a diode;

- define diode breakdown;

- differentiate between zeners and other diodes;

- determine when a diode can be considered "perfect."

1. Both silicon and germanium are semiconductor materials, and both are used in the manufacture of diodes, transistors, and other components. Both are refined to an extreme level of purity, then minute, controlled amounts of a specific impurity are added. Depending on which impurity

is added, the silicon or germanium is said to be N or P material. The electric current in semiconductor material is made up of two types of charge carriers. In N material the majority carriers are negative in charge (electrons), and the minority carriers are positive in charge (called holes). In P material the opposite is true. It is not necessary to go into the physics of these materials in any more detail.

When a piece of N silicon and a piece of P silicon are joined together, a *diode junction*, often called a PN junction, is formed. Diode junctions can also be made with N and P germanium. However, silicon and germanium are never mixed when making PN junctions. Which

diagrams below show diode junctions? _____

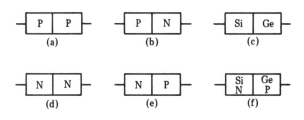

diagrams b and e only

2. In a diode the P material is called the *anode*. The N material is called the *cathode*. In the figure below, label the diode segments as P or N.

The anode is labeled P, the cathode N.

3. The main characteristic of the diode is that current will flow through it *in one direction only*. The figure below shows the direction in which the current flows.

In circuits the diode is given the symbol shown below. The arrowhead points in the direction of current flow. The anode and cathode are indicated here, but are not usually in circuit diagrams. Therefore you must know which is which.

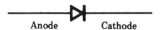

Does current flow from anode to cathode, or cathode to anode in a diode? _____

‒ ‒ ‒ ‒ ‒ ‒ ‒ ‒ ‒ ‒ ‒ ‒ ‒ ‒ ‒ ‒ ‒

flows from anode to cathode

4. In the circuit shown below, an arrow shows the assumed current flow.

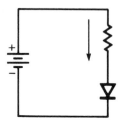

   (a) Is the diode connected correctly to permit current to flow? _____

   (b) Notice the way the battery and the diode are connected. Is the anode at a higher or lower voltage than the cathode? _____

‒ ‒ ‒ ‒ ‒ ‒ ‒ ‒ ‒ ‒ ‒ ‒ ‒ ‒ ‒ ‒ ‒

(a) yes; (b) higher

5. When the diode is connected so that the current is flowing, it is said to be *forward biased.* In a forward biased diode, the anode is connected to a higher voltage than the cathode, and current is flowing. Examine the way the diode is connected to the battery in the figure below.

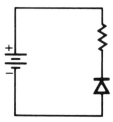

Is the diode forward biased or not? Give the reasons for your choice.

_____

‒ ‒ ‒ ‒ ‒ ‒ ‒ ‒ ‒ ‒ ‒ ‒ ‒ ‒ ‒ ‒

No, it is not forward biased. The cathode is connected to the higher voltage and no current can flow because the diode can only conduct current flow in the forward direction.

6. When the cathode is connected to a higher voltage level than the anode,

the diode cannot conduct. In this case the diode is said to be *back biased*, or *reverse biased.*

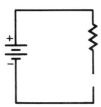

Draw in a back biased diode in this circuit.

_ _ _ _ _ _ _ _ _ _ _ _ _ _ _ _

Your drawing should look something like this.

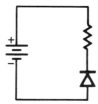

7. In many circuits the diode is often considered to be a *perfect* diode to simplify calculations. A perfect diode has zero voltage drop in the forward direction and conducts no current in the reverse direction.

    From your basic electricity, what other component has zero voltage drop across its terminals in one condition, and conducts no

    current in an alternative condition? _____

    _ _ _ _ _ _ _ _ _ _ _ _ _ _ _ _

    The mechanical switch. When closed it has no voltage drop across its terminals, and when open it conducts no current.

8. A forward biased perfect diode can thus be compared to a closed switch. It has no voltage drop across its terminals and current flows through it.

    A reverse biased diode can be compared to an open switch. No current flows through it and it will have the full circuit voltage appear across its terminals.

    Which of the switches below could be replaced by a forward biased

    perfect diode? _____

_ _ _ _ _ _ _ _ _ _ _ _ _ _ _ _

switch 2

THE DIODE EXPERIMENT

9.   If you have access to electronic equipment you may wish to perform
the simple experiment described in the next few frames. If this is the
first time you have tried such an experiment, get help from an instruc-
tor or someone who is familiar with electronic experiments.

If you do not have access to equipment do not skip this frame.
Read through the experiment and try to picture or imagine what will be
happening. This is sometimes called "dry-labbing" the experiment and
a lot can be learned from it, even though it is always better to actually
perform the experiment. This advice also applies to the other experiments that are given in many of the following chapters.

The object of the experiment is to plot the V-I curve or characteristic curve of the diode, to show how the current flow through the
diode varies with the applied voltage. Characteristic curves have several
uses in electronics, and this experiment introduces one of the most
simple and fundamental of these curves. In Chapter 1 a V-I curve for
the resistor showed a straight line (see frame 22). The characteristic
curve for the diode is not a straight line. It is shown below.

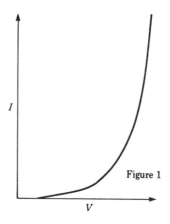

Figure 1

In performing this experiment you will gain experience in:

(1)   setting up a simple electronic experiment;
(2)   measuring voltage and current;
(3)   plotting a graph of these.

Set up the circuit shown on the next page. The circled A and V desig-
nate meters. The ammeter will measure current, and the voltmeter will
measure voltage in the circuit.

You can use ordinary current meters and voltmeters, a test meter,
or a multimeter. The variable resistor can be a substitution box, various
individual resistors, or a potentiometer. The highest value of resistance
needed will be 1 megohm. Carefully check your circuit against the diagram, especially the direction of the battery and the diode.

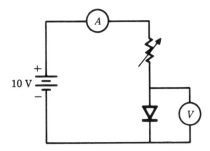

Once you have checked your circuit, follow this procedure.

(1) Set R to its highest value and record it in the table.
(2) Close the switch and measure I and V. Record them in the table.
(3) Lower the value of R a little to get a different reading of I.
(4) Measure and record I and V again.
(5) Continue in this fashion, taking as many readings as possible. There will suddenly come a point when V will not increase, but I will increase very rapidly. STOP.
    Note: If V gets very large—above 3 or 4 volts—and I remains very small, then the diode is backwards. Reverse it and start again.
(6) Graph the points recorded in the table, using the blank graph on page 32. Your curve should look like the one in figure 1.

R ohms	V volts	I mA

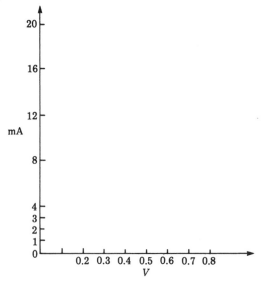

10. The measurements in the following table were taken using a commercial
     1N4001 diode.

R ohms	V volts	I mA
1 MΩ	0.30	0.02
220 kΩ	0.40	0.05
68	0.46	0.14
33	0.50	0.26
15	0.52	0.50
10	0.55	0.80
6.8	0.56	1.20
4.7	0.60	2.00
3.3	0.62	2.80
2.2	0.64	4.20
1.5	0.65	5.50
1.0	0.67	8.40
680 Ω	0.70	12.00
470	0.70	18.00
330	0.71	23.00

Further reductions in the value of R will cause very little increase in the
voltage but will produce large increases in the current.

This is the *V-I* curve for the measurements shown in the table.

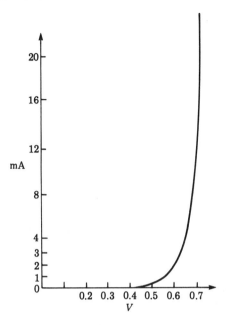

The *V-I* curve, or diode characteristic curve, is repeated here with three important regions marked on it.

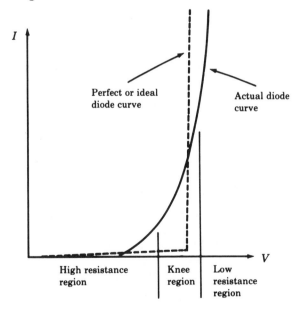

    The most important region is the *knee region.* This is not a sharply defined changeover point, but it occupies a very narrow range of the curve, where the diode resistance changes from high to low.
    The ideal curve is shown for comparison.

For the diode used in this frame the knee voltage is about 0.7 V, which is typical for a silicon diode. This means, and your data should verify this, at voltage levels below 0.7 V, the diode has such a high resistance that it limits the current flow to a very low value. This characteristic knee voltage is sometimes referred to as a *threshold voltage*. If you used a germanium diode the knee voltage is about 0.3 V.

What is the knee voltage for the diode you used? _____

_ _ _ _ _ _ _ _ _ _ _ _ _ _ _ _

If you used silicon, the knee voltage would be approximately 0.7 V; if germanium, approximately 0.3 V.

11.   The knee voltage is also a limiting voltage. That is, it is the highest voltage that can be obtained across the diode in the forward direction.

(a)   Which has the higher limiting voltage, germanium or silicon?

_____

(b)   What happens to the diode resistance at the limiting or knee

voltage? _____

_ _ _ _ _ _ _ _ _ _ _ _ _ _ _ _ _

(a)   Silicon, with a limiting voltage of 0.7 V, is higher than germanium with a limiting voltage of only 0.3 V.
(b)   It changes from high to low.

Note: We will be using these knee voltages in many later chapters as the voltage drop across the PN junction when it is forward biased.

12.   Refer back to the diagram of resistance regions in frame 10. What happens to the current when the voltage becomes limited at the knee?

_____

_ _ _ _ _ _ _ _ _ _ _ _ _ _ _ _

It increases rapidly.

13.   For any given diode the knee voltage will not be exactly 0.7 V or 0.3 V. It will vary slightly. But when using diodes in practice—that is, imperfect diodes—two assumptions can be made.

(1)   The voltage drop across the diode is either 0.7 V or 0.3 V.
(2)   Excessive current is prevented from flowing through the diode by a suitable choice of resistor in series with the diode.

(a)   Why do we specify imperfect diodes here? _____

(b) Would you use a high or low resistance to prevent excessive

current? _____

— — — — — — — — — — — — — — — —

(a) All diodes are imperfect and the 0.3 or 0.7 voltage values are only
approximated. In fact, we will be making an additional
assumption later on that the voltage drop across the diode, when
it is conducting, is 0 V. This assumes, then, that the ideal
characteristic curve indicates low resistance (in this case 0 ohms)
immediately upon seeing a small voltage and that the knee is at
0 V.
(b) Generally a high resistance; however, its actual value depends on
the applied voltage and the maximum current the diode can stand.

14. Calculate the current through the diode in this circuit, using the steps
below.

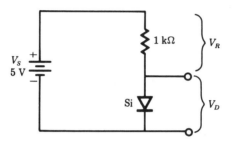

(a) The voltage drop across the diode is known.
It is 0.7 V for silicon and 0.3 V for germanium. ("Si" near the diode means it is
silicon.) Write down the diode voltage
drop. $V_D =$ _____

(b) Find the voltage drop across the resistor.
This is given by $V_R = V_S - V_D$. (V_S is the
supply voltage, or battery voltage.)
This is taken from KVL, which was
discussed in Chapter 1. $V_R =$ _____

(c) Calculate the current through the
resistor. Use $I = V_R /R$ $I =$ _____

(d) Finally, determine the current through
the diode. $I_D =$ _____

— — — — — — — — — — — — — — — —

You should have written these values.

(a) 0.7 V
(b) $V_R = V_S - V_D = 5 \text{ V} - 0.7 \text{ V} = 4.3 \text{ V}$

(c) $I = \dfrac{V_R}{R} = \dfrac{4.3 \text{ V}}{1 \text{ k}\Omega} = 4.3 \text{ mA}$

(d) 4.3 mA

15. In practice, when the battery voltage is 10 V or above, the voltage drop across the diode is often considered to be 0 V instead of 0.7 V.

    The assumption made here is that the diode is a perfect diode and the knee voltage is at 0 V rather than a threshold value that must be exceeded. As discussed later, this assumption is often used in many electronic design situations.

(a) Calculate the current through the silicon diode in this circuit.

$V_D = $ _____

$V_R = V_S - V_D = $ _____

$I = \dfrac{V_R}{R} = $ _____

$I_D = $ _____

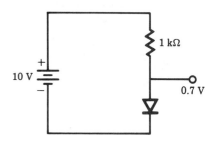

(b) Calculate the current through the perfect diode in this circuit.

$V_D = $ _____

$V_R = V_S - V_D = $ _____

$I = \dfrac{V_R}{R} = $ _____

$I_D = $ _____

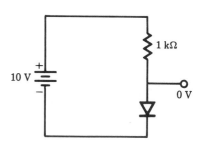

- - - - - - - - - - - - - - - -

(a) 0.7 V; 9.3 V; 9.3 mA; 9.3 mA
(b) 0 V; 10 V; 10 mA; 10 mA

16. The difference in the values of the two currents just found is less than 10% of the total current. That is, 0.7 mA is less than 10% of 10 mA. Most electronic components have more than 10% tolerance in their nominal values. This means that a 1 k resistor can be anywhere from 900 ohms to 1100 ohms and still be valid. Hence the value of current through a resistor in practice can be 10% different from that calculated, and can change by 10% if the resistor is changed.

    Thus calculations are often simplified if the simplification does not change values by more than 10%. A diode is often assumed to be perfect when the circuit voltage is 10 V or more.

(a)   Examine the circuit below. Determine if it is safe to consider the
       diode perfect or not. _____

(b)   Calculate the current through the diode. _____

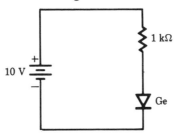

(a)   It can be considered a perfect diode.
(b) $I = 10$ mA

17.   When a current flows through a diode it causes heating and power dis-
       sipation, just as with a resistor. The power formula for resistors is
       $P = V \times I$. This same formula can be applied to diodes to find the power
       dissipation. To calculate the power dissipation in a diode, you must
       first calculate the current as shown previously. The voltage drop in this
       formula is assumed to be 0.7 V for a silicon diode, even if you con-
       sidered it to be 0 V when calculating the current.
              For example: A diode has 100 mA flowing through it. Find the
       power it dissipates.

$$P = (0.7)(100 \text{ mA}) = 70 \text{ mW}$$

       Assume a current of 2 amperes is flowing through a diode. How

       much power is being dissipated? _____

$P = (0.7 \text{ V})(2 \text{ A}) = 1.4$ watts

18.   Diodes are made to dissipate a certain amount of power, and this is
       quoted as a maximum power rating in the specifications of the diode by
       the manufacturer.
              Assume a diode has a maximum power rating of 2 watts. How
       much current can it safely pass?

$$P = V \times I$$

$$I = \frac{P}{V}$$

$$= \frac{2 \text{ watts}}{0.7 \text{ V}}$$

$$= 2.86 \text{ A (rounded off to two decimal places)}$$

Provided the current in the circuit does not exceed this, the diode will not overheat and burn out.

Suppose the maximum power rating of a germanium diode is 3 watts. What is its highest safe current? _____

— — — — — — — — — — — — — — —

$$I = \frac{3 \text{ watts}}{0.3 \text{ V}} = 10 \text{ A}$$

19. (a) Would a 3 watt silicon diode be able to carry the current calculated for the germanium diode for frame 18? _____

_____

(b) What would be its safe current? _____

— — — — — — — — — — — — — — —

(a) No, 10 amperes would cause a power dissipation of 7 watts, which would burn up the diode.

(b) $I = \frac{3}{0.7} = 4.3 \text{ A}$

Any current less than this would be safe.

20. In the next several examples we are going to concentrate on finding the current through the diode. Look at the circuit shown below.

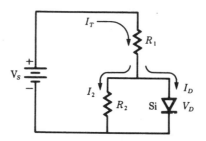

The total current from the battery flows through R_1, and then splits into I_2 and I_D. I_2 flows through R_2 and I_D through the diode.

(a) What is the relationship between I_T, I_2, and I_D? _____

(b) What is the value of V_D? _____

— — — — — — — — — — — — — — —

(a) Remember KCL, $I_T = I_2 + I_D$

(b) $V_D = 0.7 \text{ V}$

21. To find I_D it is necessary to go through the steps below, as there is no way of finding I_D directly.

(a) Find I_2. This is done using $V_D = R_2 \times I_2$.

(b) Find V_R. For this use $V_R = V_S - V_D$, KVL again.

(c) Find I_T, the current through R_1. Use $V_R = I_T \times R_1$.

(d) Find I_D. This is found by using $I_T = I_2 + I_D$, KCL again.

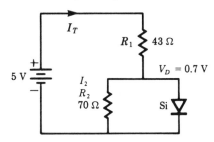

To find I_D in the circuit here, go through the above steps, then check your answers.

(a) $I_2 =$ _____

(b) $V_R =$ _____

(c) $I_T =$ _____

(d) $I_D =$ _____

- - - - - - - - - - - - - - - -

(a) $I_2 = \dfrac{V_D}{R_2} = \dfrac{0.7 \text{ V}}{70 \text{ ohms}} = 0.01 \text{ A} = 10 \text{ mA}$

(b) $V_R = V_S - V_D = 5 \text{ V} - 0.7 \text{ V} = 4.3 \text{ V}$

(c) $I_T = \dfrac{V_R}{R_1} = \dfrac{4.3 \text{ V}}{43 \text{ ohms}} = 0.1 \text{ A} = 100 \text{ mA}$

(d) $I_D = I_T - I_2 = 100 \text{ mA} - 10 \text{ mA} = 90 \text{ mA}$

22. What is the power dissipation of the diode in frame 21?

_____

- - - - - - - - - - - - - - -

$P = V_D \times I_D = (0.7 \text{ V})(90 \text{ mA}) = 63 \text{ milliwatts}$

23. Now find the current in the diode for this circuit. Fill in the steps shown.

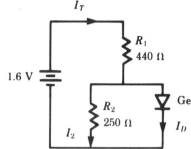

(a) $I_2 =$ _____

(b) $V_R =$ _____

(c) $I_T =$ _____

(d) $I_D =$ _____

- - - - - - - - - - - - - - - - - -

(a) $I_2 = \dfrac{0.3}{250} = 1.2 \text{ mA}$

(b) $V_R = V_S - V_D = 1.6 - 0.3 = 1.3 \text{ V}$

(c) $I_T = \dfrac{V_R}{R_1} = \dfrac{1.3}{440} = 3 \text{ mA}$

(d) $I_D = I_T - I_2 = 1.8 \text{ mA}$

If you are going to take a break soon, this is a good stopping point.

DIODE BREAKDOWN

24. We mentioned earlier that if the experiment is not working correctly, then the diode is probably in backwards. If the diode is placed in the circuit backwards—as in the diagram on the right below—then almost no current flows. In fact, the current flow is so small, we can say that no current flows. The *V–I* curve for a reversed diode looks like the one shown below.

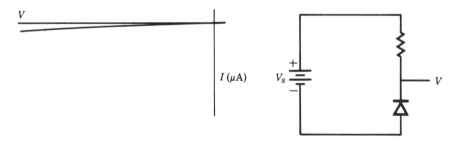

In theory this line will go on and on. But in practice a voltage is reached where the diode "breaks down" and suddenly it stays at this voltage while a large current flows. The graph for the diode breakdown would now look like the one at the top of page 41.

If this condition is allowed to persist, the current will become excessive and the diode will burn out. A diode can be prevented from burning out, even though it is at the breakdown voltage, if the current is limited by a resistance.

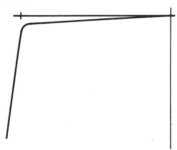

Suppose a diode is set up in the circuit shown below. It is known to break down at 100 V, and it can safely pass 1 ampere without overheating. Find the resistance in this circuit that would limit the current

to 1 ampere. _____

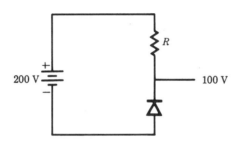

$V_R = V_S - V_D = 200\ V - 100\ V = 100\ V$

Since 1 ampere of current is flowing, then $R = V_R / I = \dfrac{100\ V}{1\ A} = 100$ ohms.

25. All diodes will break down when connected in the reverse direction if excess voltage is applied to them. The breakdown voltage, which is a function of how the diode is made, varies from one type of diode to another; it is quoted on the specification sheet published by the manufacturer.

   Breakdown is not a catastrophic process. That is, it does not destroy the diode. If the excessive supply voltage is removed, the diode will recover to normal operation. It can be used safely many times again provided the current is limited to prevent the diode burning out.

   A diode will always break down at the same voltage, no matter how many times it is used.

   The breakdown voltage is often called the *peak inverse voltage* (PIV) or the *peak reverse voltage* (PRV). Following are the PIVs of some common diodes.

Diode	PIV
1N4001	50 V
1N4002	100 V
1N4003	200 V
1N4004	400 V
1N4005	600 V
1N4006	800 V

Note: These are high voltages and do not tend to be precise.

(a) Which can permanently destroy a diode, excessive current or excessive voltage? _____

(b) Which is more harmful to a diode, breakdown or burnout?

_____

– – – – – – – – – – – – – – – – –

(a) Excessive current. Excessive voltage will not harm it if the current is limited.
(b) Burnout. Breakdown is not necessarily harmful, especially if the current is limited.

THE ZENER DIODE

26. Diodes can be manufactured so that breakdown occurs at much lower and more precise voltages than those just discussed. These types of diodes are called *zener diodes*, so named because they exhibit the "Zener effect"—a particular form of voltage breakdown. At the zener voltage a small current will flow through the zener diode. This current must be maintained in order to keep the diode at the zener point. In most cases, a few milliamperes are all that is required. The zener diode symbol and a simple circuit are shown below.

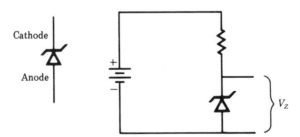

In this circuit, the battery determines the applied voltage. The zener diode determines the voltage drop, labeled V_Z, across it. The resistor determines the current flow. Zeners are used to maintain a constant voltage at some point in a circuit.

Why are zeners used for this purpose, while ordinary diodes are

not? _____

_ _ _ _ _ _ _ _ _ _ _ _ _ _ _ _ _ _

because zeners have a precise breakdown voltage

27. Let's examine an application in which a constant voltage may be desirable. Consider a lamp driven by a DC generator. When the generator is
turning at full speed it puts out 50 V. When it is running slower the
voltage can drop to 35 V. We wish to illuminate a 20 V lamp with this
generator. Assume the lamp draws 1.5 A. The circuit is shown below.

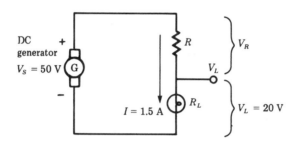

We need to determine a suitable value for the resistance. These steps will
find a suitable resistance value.

(a) Find R_L, the lamp resistance. Use $R_L = \dfrac{V_L}{I}$

(b) Find V_R. Use $V_S = V_R + V_L$

(c) Find R. Use $R = \dfrac{V_R}{I}$

Work through these steps, and write your answers below.

(a) $R_L = $ _____

(b) $V_R = $ _____

(c) $R = $ _____

_ _ _ _ _ _ _ _ _ _ _ _ _ _ _ _ _

(a) $R_L = \dfrac{20 \text{ V}}{1.5 \text{ A}} = 13.33$ ohms

(b) $V_R = 50 \text{ V} - 20 \text{ V} = 30 \text{ V}$

(c) $R = \dfrac{50 \text{ V} - 20 \text{ V}}{1.5 \text{ A}} = \dfrac{30 \text{ V}}{1.5 \text{ A}} = 20$ ohms

28. Let us assume now that the 20 ohm resistor calculated in frame 27 is in place and the voltage output of the generator drops to 35 V. This is similar to the situation where a battery gets old; its voltage level decays and it will no longer have sufficient voltage to produce the proper current. This will result in the lamp glowing less brightly, or perhaps not at all. We should note here, however, that the resistance of the lamp stays the same.

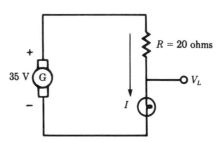

(a) Find the total current flowing. Use $I_T = \dfrac{V_S}{(R + R_L)}$

$I_T =$ _____

(b) Find the voltage drop across the lamp. Use $V_L = I_T \times R_L$

$V_L =$ _____

(c) Have the voltage and current increased or decreased? _____

- - - - - - - - - - - - - - -

(a) $I_T = \dfrac{35\ \text{V}}{20\ \Omega + 13.3\ \Omega} = \dfrac{35\ \text{V}}{33.3\ \Omega} = 1.05\ \text{A}$

(b) $V_L = 1.05\ \text{A} \times 13.3\ \Omega = 14\ \text{V}$
(c) Both have reduced in value.

29. In many applications a lowering of voltage across the lamp—or some other component—may be intolerable. It can be prevented by using a zener diode, as in this circuit.

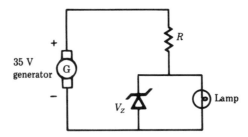

If we choose a 20 V zener—that is, one which has a 20 V drop across it—then the lamp will always have 20 V across it, no matter what the output voltage is from the generator (provided, of course, that the output from the generator is always above 20 V).

If the voltage across the lamp is constant, and the generator output drops:

(a) what happens to the current through the lamp? _____

(b) what happens to the current through the zener? _____

— — — — — — — — — — — — — —

(a) It stays constant because the voltage across the lamp stays constant.
(b) It decreases because the total current decreases.

30. In order to make this circuit work and keep 20 V across the lamp at all times, we must find a suitable value of R which will allow sufficient total current to flow to provide the 1.5 amperes drawn by the lamp and the small amount required by the diode to keep it at its zener voltage. To do this we start at the "worst case" condition. ("Worst case" design is a common practice in electronics. It is used to ensure that equipment will work under the most adverse conditions). The worst case here is when the generator is putting out only 35 V.

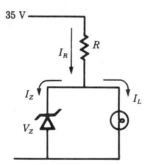

We must find the value of R which will allow 1.5 A to flow through the lamp. How much current will flow through the zener diode? We can choose any current we like, provided it is above a few milliamperes, and provided it will not cause the zener diode to burn out. In this example we will assume that the zener current I_Z is 0.5 A.

(a) What is the total current through R?

$I_R =$ _____

(b) Calculate the value of R.

$R =$ _____

— — — — — — — — — — — — — —

(a) $I_R = I_L + I_Z = 1.5\ A + 0.5\ A = 2\ A$

(b) $R = \dfrac{(V_S - V_Z)}{I_R} = \dfrac{(35\ \text{volts} - 20\ \text{volts})}{2\ A} = 7.5\ \text{ohms}$

Note that a different value of resistor was produced from that in the previous example, in frame 25. A different choice of I_Z here would produce another value of R.

31. Now let's see what happens when the generator is supplying 50 V.

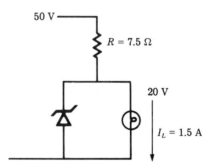

As the lamp still has 20 V across it, it will still draw only 1.5 A. But the total current and the zener current will change.

(a) Find the total current through R.

$I_R = $ _____

(b) Find the zener current.

$I_Z = $ _____

— — — — — — — — — — — — — — —

(a) $I_R = \dfrac{(V_S - V_Z)}{R} = \dfrac{(50 - 20)}{7.5} = 4$ A

(b) $I_Z = I_R - I_L = 4 - 1.5 = 2.5$ A

32. Although the lamp voltage and current have remained the same as the conditions which we wanted, the total current and the zener current have both changed.

(a) What has happened to I_T—or I_R ? _____

(b) What has happened to I_Z ? _____

— — — — — — — — — — — — — —

(a) I_T has increased by 2 A.
(b) I_Z has increased by 2 A.

Note that the increase in I_T flows through the zener diode and not through the lamp.

33. How much power is dissipated by the zener diode in each of these cases?

   (a) Find the power dissipated when the generator voltage is 35 V.

   _____

   (b) Now find the power when the generator is at 50 V.

   _____

   — — — — — — — — — — — — — — — —

   (a) $P_Z = V \times I = (20 \text{ volts}) (0.5 \text{ amperes}) = 10 \text{ watts}$
   (b) $P_Z = V \times I = (20 \text{ volts}) (2.5 \text{ amperes}) = 50 \text{ watts}$
       If a zener diode with a power rating of 50 watts or more is used, it
       will not burn out.

34. For the circuit shown, what power rating should the zener diode have?
    The rating of the lamp is given.

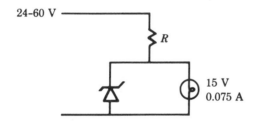

   — — — — — — — — — — — — — — — —

   At 24 volts, assume a zener current of 0.5 A.
   Then $R = \dfrac{9}{0.575} = 15.7$ ohms.
   At 60 V, $I_R = \dfrac{45}{15.7} = 2.87$ A and $I_Z =$ about 2.8 A
   $P_Z = 42$ watts

SUMMARY

Semiconductor diodes are used extensively in modern electronic circuits.
Their main advantages are listed below.

- They are very small in size.
- They are rugged and reliable if properly used. You must remember that
  excessive reverse voltage or excessive forward current could damage
  or destroy the diode.

- Diodes are very easy to use, as there are only two connections to make.

- They are inexpensive.

- They can be used in all types of electronic circuits, from simple DC controls to radio and TV circuits.

- They can be made in many sizes and can handle a wide range of current and power.

- Specialized diodes—which have not been covered here—will perform particular functions which cannot be accomplished with any other components.

- Finally, as you will see in the next chapter, diodes are an integral part of transistors.

All of the many uses of semiconductor diodes are based on the fact they conduct in *one direction only.* Typical uses for diodes are:

- logic and decision circuits in computers and control circuits;

- simple switching circuits which have no moving parts;

- converting AC to DC;

- recovering the TV and radio signals from those transmitted over the air, thus enabling us to see and hear the program.

SELF-TEST

The questions below will test your understanding of this chapter. Use a separate sheet of paper for your diagrams or calculations. Compare your answers with the answers provided following the test.

1. Draw the circuit symbol for the diode, label each terminal, and use an arrow to show the direction of current flow.

2. What semiconductor materials are used in the manufacture of diodes?

_____

3. Draw a circuit with a battery, resistor, and a forward biased diode.

4. What is meant by a reverse biased diode? _____

5. Draw a typical *V-I* curve of a forward biased diode. Show the knee voltage.

6. What is the knee voltage for silicon? _____ Germanium?

_____

7. In the circuit shown here, $V_S = 10$ V and $R = 100$ ohms. Find the current through the diode, assuming a perfect diode.

_____

8. Repeat question 7 with these values: $V_S = 3$ V and $R = 1$ kΩ.

_____

9. In this circuit find the current through the diode Si.

$V_S = 10$ V
$R_1 = 10$ kΩ
$R_2 = 1$ kΩ

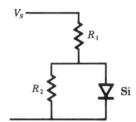

10. In this circuit find the current through the zener diode.

$V_S = 20$ V
$V_Z = 10$ V
$R_1 = 1$ kΩ
$R_2 = 2$ kΩ

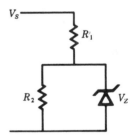

_____

11. If it is possible for the supply voltage of problem 10 to rise to 45 V, what will be the current in the zener diode? _____

_____

12. What will be the maximum power dissipated for the diode in problems 10 and 11? _____

Answers to Self-Test

If your answers do not agree with those given below, review the frames indicated in parentheses before you go on to the next chapter.

1.   Cathode ⊣◁⊢ Anode   (frame 3)

2.   germanium and silicon   (frame 1)

3.
                                                          (frame 4)

4.   one which is not conducting   (frame 6)

5.
                                                          (frame 9)

6.   Si = 0.7 V; Ge = 0.3 V (These are approximate.)   (frame 10)

7. $I_D = 100$ mA   (frame 14)

8.   As $V_S = 3$ V, do not ignore the voltage drop across the diode. Thus,
     $I_D = 2.7$ mA   (frame 14)

9.   Ignore V_D in this case. Thus, $I_D = 0.3$ mA. If V_D is not ignored,
     $I_D = 0.23$ mA.   (frame 21)

10. $I_Z = 5$ mA   (frame 30)

11. $I_Z = 30$ mA   (frame 31)

12. It will be when I_Z is at its peak value of 30 mA.
     $P_Z(\text{MAX}) = 0.35$ W   (frame 33)

CHAPTER THREE

Introduction to the Transistor

The transistor is undoubtedly the most important modern electronic component. It has made great and profound changes in electronics and in our daily lives since its discovery in 1948.

In this chapter the transistor will be introduced as an electronic component which acts similarly to a simple mechanical switch, since it is actually used as a switch in much modern electronic equipment. The transistor can be made to conduct or not conduct an electric current—which is exactly what a mechanical switch does.

An experiment in this chapter will help you to build a simple one-transistor switching circuit. This circuit can be easily set up on a home workbench, and it will enhance your learning if you obtain the few components required and actually perform the experiment of building and operating the circuit.

In the next chapter you will continue the study of switching designs and the operation of the transistor as a switch. In a much later chapter you will learn how a transistor can be made to operate as an amplifier. In this mode, the transistor reproduces an output that is a magnified version of an input signal. Rather than using the transistor to change from one state to another (ON or OFF), you can use it to increase voltage, current, or power levels. Amplification is required in many electronic circuits. These chapters taken together present an easy introduction as to why transistors are used and how they are applied in basic designs found in computers and other modern circuits.

The most common transistor type is the bipolar junction transistor, commonly called the BJT. However, it is not the only transistor type; another type will also be introduced in this chapter. This other transistor is the junction field effect transistor, or JFET. This chapter and later chapters will concentrate on the basic operation of the BJT, since it is the most common and the easiest to follow.

When you complete this chapter you will be able to:

- describe the basic construction of a bipolar junction transistor (BJT);
- specify what transistor switching action is;
- differentiate between the two most common types of transistor;
- tell which currents flow through a transistor;
- specify the relationship between base and collector current in a transistor;
- calculate the current gain for a transistor;
- explain how a transistor can be ON or OFF;
- compare the transistor to a simple mechanical switch;
- do simple transistor current calculations;
- describe the basic construction of a field effect transistor (JFET).

1. The following diagrams show the packaging techniques for several common transistor case designs in use today. For each transistor, the diagrams show the lead designations and how to identify them based on the package design. Transistors are designed to be either soldered directly into a circuit or inserted into sockets that are designed especially for them. If soldering, you must take great care, as transistors can be damaged if either the leads or case are overheated. Proper heat sinking must be used whenever soldering to transistor leads. Connections, however, may be soldered to socket terminals before a transistor is inserted, to offer complete protection from heat sources.

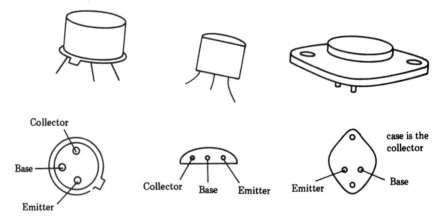

(a) How many leads are there on most transistors? _____

(b) Where there are only two leads, what takes the place of the third lead? _____

(c)   What are the three leads or connections called?

_____

(d)   Why should great care be exercised when soldering transistors into a

circuit? _____

— — — — — — — — — — — — — — — —

(a)   three
(b)   The case can be used instead, as in the diagram on the right. (This is
limited to power transistors.)
(c)   emitter, base, and collector
(d)   excessive heat can damage a transistor

2.   In its simplest form a transistor can be considered as two diodes, connected back to
back, as in this drawing.

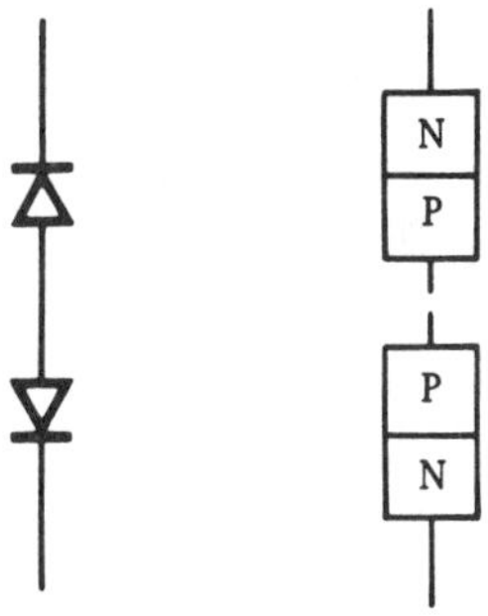

However, in the construction process one
very important modification is made. Instead
of two separate P regions as shown, only
one very thin region is used.

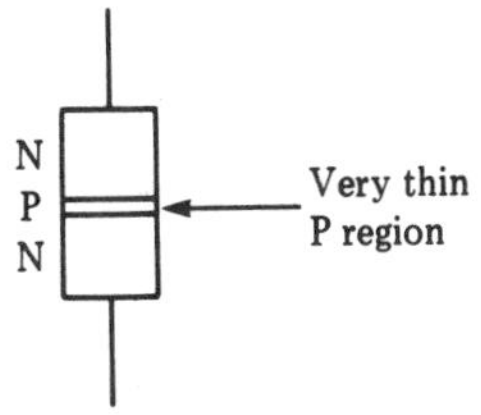

Which has the thicker P region, the transistor shown above or two

diodes connected back to back? _____

— — — — — — — — — — — — — — — —

Two diodes. The transistor has a very thin P region.

3.   Two separate diodes wired back to back will not behave like a transistor.
Why this difference in construction should make the transistor act like a
transistor and not like two diodes will not be covered, as this is semiconductor physics and not electronics.
What is the main construction difference between two diodes con-

nected back to back and a transistor? _____

_____

— — — — — — — — — — — — — — — —

the very thin P region used in the transistor

4. The three terminals of a transistor—the base, the emitter, and the collector—are connected as shown below.

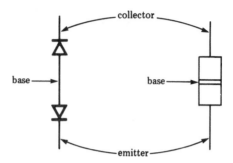

The two diodes are usually called the *base-emitter diode*, and the *base-collector diode*.

The symbol used in circuit diagrams for the transistor is shown on the diagram below, with the two diodes and the junctions shown for comparison.

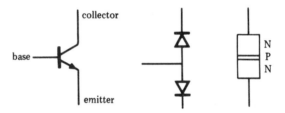

Because of the way the semiconductor materials are arranged, this is known as an NPN transistor. Which transistor terminal has the arrowhead? _____

– – – – – – – – – – – – – – – – – – –

the emitter

5. It is also possible to make transistors with a PNP configuration, as shown below.

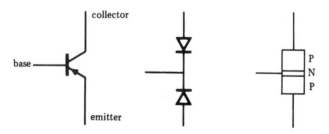

Both types, NPN and PNP, are made from either silicon or germanium.

(a)   Draw a circuit symbol for both an NPN and a PNP transistor. (Use a separate sheet of paper for your diagrams.)

(b)   Which of these might be silicon? _____

(c)   Are silicon and germanium ever combined in a transistor? (Hint: What was said about diodes?) _____

— — — — — — — — — — — — — —

(a)

NPN   PNP

(b)   Either or both could be silicon. (Either or both could also be germanium.)
(c)   Silicon and germanium are *never* mixed in any semiconductor.

6.   For simplicity of explanation we will continue with the NPN for a while. We will introduce the PNP again later.
      If a battery is connected as shown below to an NPN transistor, then a current will flow as shown.

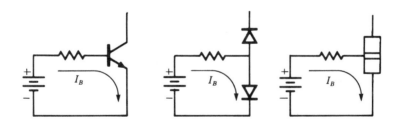

This current, flowing through the base-emitter diode, is called *base current* and is given the symbol I_B.
      Would base current flow if the battery were reversed? Give a

reason for your answer. _____

_____

— — — — — — — — — — — — — —

Base current would not flow as the diode would be back biased.

7.   In the circuit on page 56, the base current can be calculated using the techniques from Chapter 2. Find the base current in this circuit. (Hint: Do not ignore the 0.7 V drop across the base-emitter diode.)

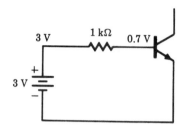

- - - - - - - - - - - - - - - - - -

Your calculations should look something like this.

$$I_B = \frac{(V_S - 0.7 \text{ V})}{R} = \frac{(3 - 0.7)}{1 \text{ k}\Omega} = \frac{2.3 \text{ V}}{1 \text{ k}\Omega} = 2.3 \text{ mA}$$

8. In the following circuit, as the 10 V battery is much higher than the 0.7 V diode drop, we can consider the base-emitter diode to be a perfect diode, and thus assume the voltage drop is 0 V.

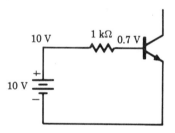

Calculate the base current.

$I_B =$ _____

- - - - - - - - - - - - - - - - - -

$$I_B = \frac{(10 - 0)}{1 \text{ k}\Omega} = \frac{10}{1 \text{ k}\Omega} = 10 \text{ mA}$$

9. Look at the circuit below.

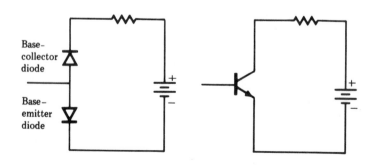

Will current flow in this circuit? Why or why not? _____

_____

_ _ _ _ _ _ _ _ _ _ _ _ _ _ _ _ _

It will not flow as the base-collector diode is reverse biased.

10. Now, we will put both of the circuits together. Note that we have two
batteries, one in each of the base and collector circuits.

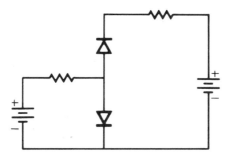

When both the base and the collector circuits, as in the preceding diagram, are connected, it demonstrates the outstanding characteristic of
the transistor, which is sometimes called *transistor action: If base current flows in a transistor, collector current will also flow.*
    Examine the current paths in the diagram below.

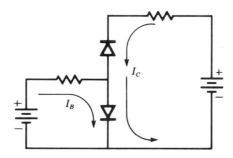

(a)   What current flows through the base-collector diode? _____

(b)   What current flows through the base-emitter diode? _____

(c)   Which of these currents causes the other to flow? _____

_ _ _ _ _ _ _ _ _ _ _ _ _ _ _ _ _

(a) I_C—the collector current
(b) I_B and I_C. Note that both of them flow through the base-emitter
      diode.
(c)   Base current causes collector current to flow.

No current flows from the collector to the base, as shown by the dotted line below.

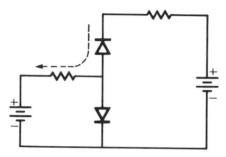

The reason why the collector current takes the path shown in frame 10, rather than the dotted line path, is beyond the scope of this book. This is the domain of semiconductor physics and is not needed in electronic circuit design and analysis at this time.

11.   Up to now we have been using the NPN transistor, solely for the purposes of illustration. A PNP transistor could have been used. There is no difference in how the two types work or behave. What is said about one is equally true for the other.

   There is, however, one important circuit difference which is illustrated below. This is caused by the fact that the PNP is made with the diodes in the reverse direction from the NPN.

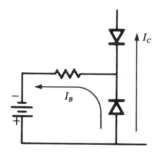 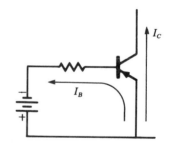

   Compare these diagrams with the second one in frame 10. How are these different in terms of the following?

(a)   Battery connections. _____

(b)   Current flow. _____

_ _ _ _ _ _ _ _ _ _ _ _ _ _ _ _ _ _

(a)   The battery is reversed in polarity.
(b)   The current flows in the opposite directions.

12. The two figures below show the battery connections to produce currents for both circuits.

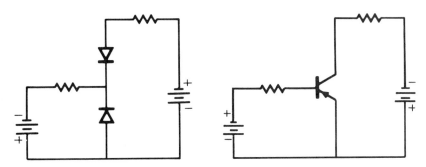

In which direction do these currents circulate? Clockwise or counter-

clockwise? _____

- - - - - - - - - - - - - - - - - -

Base current is counterclockwise.
Collector current is clockwise.

As stated earlier, there is absolutely no difference between NPN and PNP transistors. Both are used equally in electronic circuits; one is not favored over the other. Base current causes collector current to flow in both. To avoid confusion, the rest of this discussion will be conducted using NPNs only as examples. And from now on, we will use the circuit symbols only.

13. Consider the action of the circuit below. It uses only one battery to provide both the base and the collector current. The path of the base current only is shown in the diagram.

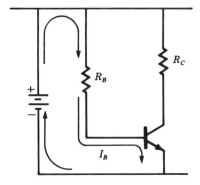

(a) Name the components through which the base current flows.

_____

(b) Into which terminal of the transistor does the base current flow?

_____

(c) Out of which transistor terminal does the base current flow?

_____

(d) Through which terminals of the transistor does no base current

flow? _____

– – – – – – – – – – – – – – – –

(a) the battery, the resistor R_B, and the transistor
(b) base
(c) emitter
(d) collector

14. Can you remember the outstanding physical characteristic of the transistor? When base current flows in the preceding circuit, what other current will flow, and which components will it flow through?

_____

_____

– – – – – – – – – – – – – – – –

Collector current will flow. It will flow through the resistor R_C and the transistor.

15. The path of the collector current is shown in the diagram below.

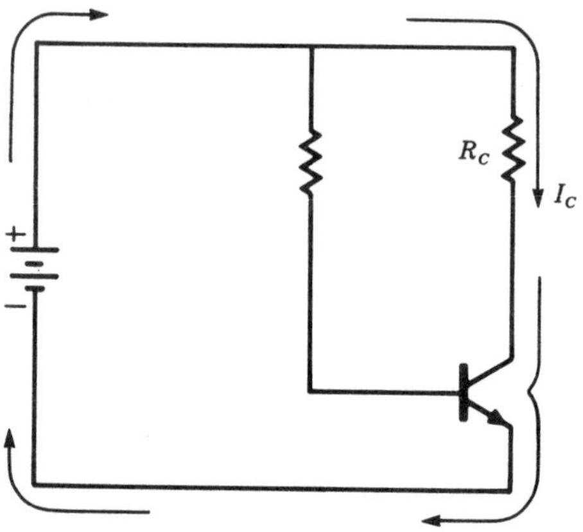

(a) List the components through which the collector current flows.

_____

(b) What causes the collector current to flow? _____

– – – – – – – – – – – – – – – –

(a) the resistor R_C, the transistor, and the battery
(b) base current (Collector current cannot ever flow if base current is not flowing first.)

16. It is a peculiar property of the transistor that the *ratio of collector current to base current is constant.* The collector current is always much larger than the base current. The constant ratio of the two currents is called the *current gain* of the transistor and it is a number much larger than 1.

Current gain is given the symbol β, called beta. Typical values of β range from 10 to 300. 100 is a good typical value from many transistors, and we will use this number for convenience in our calculations.

(a) What is the ratio of collector current to base current called?

_____

(b) What is the symbol used for this? _____

(c) Which is larger—base or collector current? _____

(d) Look back at the circuit in frame 13. Will current be greater in R_B

or in R_C? _____

- - - - - - - - - - - - - - - - - -

(a) current gain
(b) β
(c) Collector current is larger.
(d) The current is greater in R_C, since it is the collector current.

Note: The β introduced here is referred to in manufacturers' specification sheets as h_{FE}. Technically it is referred to as the static or DC β. For the purposes of this chapter we will just call it β. The h symbol refers to an h parameter. Discussions on h parameters or transistor parameters in general, of which there are many, are in textbooks, and, while important, will not be covered here. We will see another h parameter in another chapter.

17. The mathematical formula for current gain is as follows.

$$\beta = \frac{I_C}{I_B}$$

where:

I_B = base current
I_C = collector current

From this you can see that if no base current flows, no collector current will flow. And if more base current flows more collector current will flow. This is what is meant when we say the "base current controls the collector current."

Suppose the base current is 1 mA and the collector current is

150 mA. What is the current gain of the transistor? _____

- - - - - - - - - - - - - - - - -

150

18. Current gain is a physical property of the transistor. Its value can be taken from the manufacturer's published data sheets or it can be determined experimentally by the user.

In general β is a different number from one transistor type to the next, but it remains constant for a given transistor. Transistors of the same type have β values within a narrow range of each other.

One of the most often performed calculations in transistor work is determining the values of either collector or base current, when β and the other current are known.

For example, suppose a transistor has 500 mA of collector current flowing and it is known to have a β value of 100. Find the base current. To do this, use the formula as shown below.

$$\beta = \frac{I_C}{I_B}$$

$$I_B = \frac{I_C}{\beta} = \frac{500 \text{ mA}}{100} = 5 \text{ mA}$$

Now you do these simple examples.

(a) $I_C = 2$ A, $\beta = 20$. Find I_B. _____

(b) $I_B = 1$ mA, $\beta = 100$. Find I_C. _____

(c) $I_B = 10$ μA, $\beta = 250$. Find I_C. _____

(d) $I_B = 0.1$ mA, $I_C = 7.5$ mA. Find β. _____

– – – – – – – – – – – – – – – –

(a) 0.1 A, or 100 mA
(b) 100 mA
(c) 2500 μA, or 2.5 mA
(d) 75

19. This frame will serve as a summary of the first part of this chapter. You should be able to answer all the questions. Use a separate sheet of paper for your diagram and calculations.

(a) Draw a transistor circuit utilizing an NPN transistor, a base resistor, a collector resistor, and one battery to supply both base and collector currents. Show the paths of I_B and I_C.

(b) Which current controls the other? _____

(c) Which is the larger current, I_B or I_C? _____

(d) $I_B = 6$ μA, $\beta = 250$. Find I_C. _____

(e) $I_C = 300$ mA, $\beta = 50$. Find I_B. _____

– – – – – – – – – – – – – – –

(a)   Refer to frame 13 to see if the figure is correct.
(b) I_B (base current) controls I_C (collector current).
(c) I_C
(d)   1.5 mA
(e)   6 mA

THE TRANSISTOR EXPERIMENT

20.   The object of the following experiment is to find β of a particular transistor by measuring several values of base current with their corresponding values of collector current. These values of collector current will be
divided by the values of the base current to obtain β. The value of β will
be almost the same for all the measured values of current. This will
show that β is a constant for a transistor.

As long as the circuit is set up, measure the collector voltage for
each current value. This will demonstrate experimentally some points
to be made in future frames. Observe how the collector voltage V_C drops
towards 0 V as the currents are increased.

If you do not have the facilities for setting up the circuit and
measuring the values, just read through the experiment to find out how
it would be done. If you do have the facilities, you will need the following equipment and supplies.

   1 9 V transistor radio battery (or a lab power supply)
   1 current meter, maximum reading 100 μA
   1 current meter (maximum reading 10 mA)
   1 voltmeter, maximum reading 10 V
   1 resistor substitution box, or a 1 MΩ potentiometer, or
         assorted resistors with values in the table
   1 1 kΩ resistor
   1 transistor, preferably NPN

Almost any small commercially available transistor will do for this
experiment. The measurements given in this book were obtained using
a 2N3643. If only a PNP is available, then simply reverse the battery
voltage and proceed as described.

Set up the circuit shown in the following figure. Make connections
by using clip leads and take care to keep the clips from touching each
other near the transistor leads. If the connections to the transistor are
made by soldering, then refer back to frame 1 and take proper care. It
may be best to use a transistor socket or some type of plug strip. These
offer an easy method of connection to the transistor and may be
obtained from any electronics supply store.

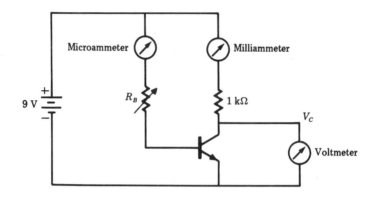

Follow this procedure.

   (1) Set R_B to its highest value.
   (2) **Measure and record I_B (in the table on page 65)**
   (3) Measure and record I_C.
   (4) **Measure and record V_C. This voltage is sometimes referred to as V_{CE}, or the collector-emitter voltage, as it is taken across the collector-emitter leads if the emitter is connected to ground or the negative of the power supply.**
   (5) Lower the value of R_B enough to produce a different reading of I_B.
   (6) Measure and record I_B, I_C, and V_C.
   (7) Lower R_B again and get a new I_B.
   (8) Measure and record all the values again.
   (9) Continue this until $V_C = 0$ V.
  (10) Further reductions in the value of R_B will increase I_B, but will not increase I_C or V_C.

Check the figures in your table to make sure you got a consistent pattern. Then compare your measurements with the ones given following the dashed line.

R_B	I_B	I_C	

The figures in this table were obtained in an experiment conducted with considerable care. Precision resistors were used, and a commercial 2N3643 transistor was used. With ordinary 10% or 20% tolerance resistors and a transistor chosen at random, different figures will obviously be obtained. So if your figures are not as precise as those here, do not worry.

R_B	I_B	I_C	V_C
1 MΩ	9 μA	0.9 mA	8.1 volts
680 kΩ	13	1.3	7.7
470	19	1.9	7.1
330	27.3	2.8	6.2
270	33.3	3.3	5.7
220	40	4.1	5.0
200	45	4.5	4.5
180	50	5	4.0
160	56	5.6	3.4
150	60	6	3
120	75	7.5	1.5
110	82	8.0	1.0
100	90	9	0.3

In the experiment which produced the table on page 65, $\beta = 100$. You can see this by forming the ratio I_C/I_B for almost every pair of current values.

For each value of I_B and its corresponding value of I_C in the experiment, calculate the value of β. The values will vary slightly but will be close to an average. (Excessively low and high values of I_B may produce values of β which will be quite different. Ignore these for now.) Did you get a consistent β? Was it close to the manufacturer's specifications for your transistor?

21. In the experiment you measured the voltage level at the collector—V_C—and recorded your measurements. Let's examine how to determine the voltage at the collector, when measurement isn't feasible. This will also show how the collector voltage can be determined without measurement.

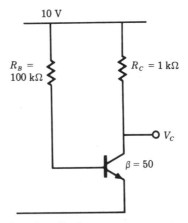

Apply these steps to the circuit above.

(a) Determine I_C.
(b) Determine the voltage drop across the collector resistor R_C. Call this V_R.
(c) Subtract V_R from the supply voltage to get the collector voltage.

Let us go through the first step together.

(a) To find I_C, we must first find I_B.

$$I_B = \frac{10 \text{ V}}{100 \text{ k}\Omega} = 0.1 \text{ mA}$$

$$I_C = \beta \times I_B = 50 \times 0.1 \text{ mA} = 5 \text{ mA}$$

Now do the next two steps. Use a separate sheet of paper for your calculations.

———————————————

(b) To find V_R:

$$V_R = R_C \times I_C = 1 \text{ k}\Omega \times 5 \text{ mA} = 5 \text{ V}$$

(c) To find V_C:
$V_C = V_S - V_R = 10\ \text{V} - 5\ \text{V} = 5\ \text{V}$

22. Use the circuit from frame 21, but use a value $\beta = 75$. Again find:

(a) I_C —————————————————————

(b) V_R —————————————————————

(c) V_C —————————————————————

- - - - - - - - - - - - - -

(a) $I_B = \dfrac{10\ \text{V}}{100\ \text{mA}} = 0.1\ \text{mA};\ I_C = 75 \times 0.1\ \text{mA} = 7.5\ \text{mA}$

(b) $V_R = 1\ \text{k}\Omega \times 7.5\ \text{mA} = 7.5\ \text{V}$
(c) $V_C = 10\ \text{V} - 7.5\ \text{V} = 2.5\ \text{V}$

23. Use the same circuit again, but with these values: $R_B = 250\ \text{k}\Omega$ and $\beta = 75$. Again find:

(a) I_C —————————————————————

(b) V_R —————————————————————

(c) V_C —————————————————————

- - - - - - - - - - - - - - - -

(a) $I_B = \dfrac{10\ \text{V}}{250\ \text{k}\Omega} = \dfrac{1}{25}\ \text{mA};\ I_C = 75 \times \dfrac{1}{25}\ \text{mA} = 3\ \text{mA}$

(b) $V_R = 1\ \text{k}\Omega \times 3\ \text{mA} = 3\ \text{V}$
(c) $V_C = 10\ \text{V} - 3\ \text{V} = 7\ \text{V}$

24. From the preceding problems you can see that V_C can be set to any desired value by choosing a transistor with an appropriate value of β, or by choosing the correct value of R_B.
Consider now this special example. The object is to find V_C. Go through the steps of the previous few examples.

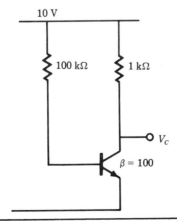

Write down your values for:

(a) $I_B =$ _____

     $I_C =$ _____

(b) $V_R =$ _____

(c) $V_C =$ _____

- - - - - - - - - - - - - - - -

You should have these values.

(a) $I_B = \dfrac{10 \text{ V}}{100 \text{ k}\Omega} = 0.1 \text{ mA}$

$I_C = 100 \times 0.1 \text{ mA} = 10 \text{ mA}$
(b) $V_R = 1 \text{ k}\Omega \times 10 \text{ mA} = 10 \text{ V}$
(c) $V_C = 10 \text{ V} - 10 \text{ V} = 0 \text{ V}.$

Here the base current is just sufficient to produce a collector voltage of
0 V. This is a very important condition with many practical applications.

25. Look at the two drawings below and compare their voltages at the point
     labeled V_C.

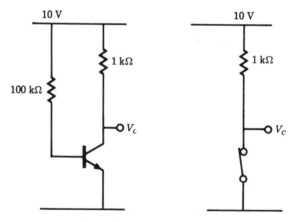

Consider a transistor which has sufficient base current and collector
current to set its collector voltage to 0 V. Obviously it can be compared
to a closed mechanical switch. As the switch is said to be ON, then the
transistor is also said to be "turned on" or just ON.

(a) From the above, what can a turned on transistor be compared to?

_____

(b) What is the collector voltage of an ON transistor? _____

- - - - - - - - - - - - - - - -

(a)   a closed mechanical switch
(b)   0 V

26.   Now compare these two circuit drawings.

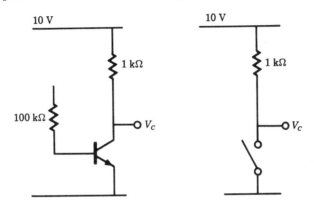

Since the base circuit is broken—that is, it is not complete—there is no base current flowing.

(a)   How much collector current is flowing? _____

(b)   What is the collector voltage? _____

(c)   What is the voltage at the point V_C in the mechanical switch

      circuit? _____

- - - - - - - - - - - - - - - - - - -

(a)   None.
(b)   Since there is no current flowing through the 1 kΩ resistor, there is no voltage drop across it. So the collector will be at 10 V.
(c)   10 V, since there is no current flowing through the 1 kΩ resistor.

27.   From frame 26, it is obvious that a transistor with no collector current can be considered similar to an open mechanical switch. For this reason a transistor with no collector current and its collector voltage level at the supply voltage level is said to be "turned off" or just OFF.
      What are the two main characteristics of an OFF transistor?

      _____

      _____

      - - - - - - - - - - - - - - -

It has no collector current, and the collector voltage is equal to the supply voltage.

28. Now work through this example and compare the results to the examples in frames 26 and 27. Again the object here is to find V_C.

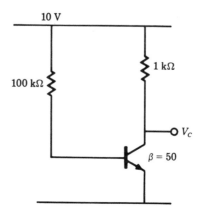

(a) I_B = _____

I_C = _____

(b) V_R = _____

(c) V_C = _____

- - - - - - - - - - - - - - - -

(a) $I_B = \dfrac{10 \text{ V}}{100 \text{ k}\Omega} = 0.1 \text{ mA}$

$I_C = 50 \times 0.1 \text{ mA} = 5 \text{ mA}$

(b) $V_R = 1 \text{ k}\Omega \times 5 \text{ mA} = 5 \text{ V}$

(c) $V_C = 10 \text{ V} - 5 \text{ V} = 5 \text{ V}$

Note the output voltage this time is exactly half the supply voltage. This condition is very important in AC electronics and will be returned to in the AC section.

THE JUNCTION FIELD EFFECT TRANSISTOR

29. Up to now the only transistor described was the bipolar junction transistor (BJT). Another transistor type that has come into use is the junction field effect transistor (JFET). The JFET, like the BJT, is used in many switching and amplification applications. The JFET is preferred when a high input impedance circuit is needed. The BJT has a relatively low input impedance as compared to the JFET. Like the BJT, the JFET is a three terminal device. The terminals are called the source, drain, and gate. They are similar in function to the emitter, collector, and base, respectively.

(a) How many terminals does a JFET have, and what are they called?

_____

(b) Which terminal has a function similar to the base of a BJT?

_____

— — — — — — — — — — — — — — — —

(a) three, called the source, drain, and gate
(b) The gate has a control function similar to that of the base of a
BJT.

30. The basic design of a JFET consists of one type of semiconductor
material with a channel made of the opposite semiconductor material
running through it. If the channel is of N material, it is called an
N-channel JFET. If it is of P material, it is called a P-channel. A
drawing showing the basic layout of N and P materials, along with
their circuit symbols, is shown here. The gate controls the current flow
through the drain and source by controlling the effective width of the
channel, allowing more or less current to flow. Thus, the voltage on the
gate acts to control the drain current just as the voltage on the base of
a BJT acts to control the collector current.

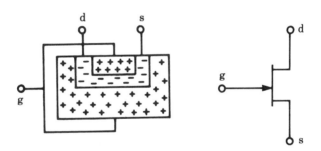

N-channel JFET

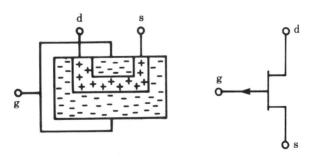

P-channel JFET

(a) Which JFET would use electrons as the primary charge carrier for the drain current? _____

(b) What effect does changing the voltage on the gate have on the operation of the JFET? _____

- - - - - - - - - - - - - - - - -

(a) N-channel, since N material uses electrons as the majority carrier.
(b) It changes the current in the drain. The channel width is controlled electrically by the gate potential.

31. To operate the N-channel JFET, a positive voltage is applied to the drain with respect to the source. This allows a current to flow through the channel. If the gate is at 0 V, the drain current will be at its largest value for safe operation and the JFET will be in the ON condition. When a negative voltage is applied to the gate, the drain current will be reduced. As the gate voltage becomes more negative, the current lessens until cutoff, which occurs when the JFET is in the OFF condition. A typical biasing circuit for the N-channel JFET is shown here. For a P-channel JFET, the polarity of the bias supplies must be reversed.

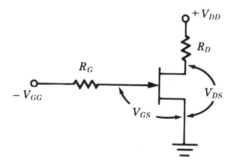

How does the ON-OFF operation of a JFET compare to that of a BJT? _____

_____

_____

- - - - - - - - - - - - - - - - -

The JFET is turned ON with 0 V on the gate, whereas the BJT is turned ON with a large voltage on the base. The JFET is turned OFF with a large voltage on the gate, whereas the BJT is turned OFF with 0 V on the base. The JFET is a "normally ON" device, but the BJT is considered a "normally OFF" device. The JFET, thus, can be utilized as a switching device.

32. For the JFET shown in frame 31: When the gate to source voltage is at 0 V ($V_{GS} = 0$), the drain current will be at its maximum or saturation value. This means that the N-channel resistance will be at its lowest possible value, in the range of 5 to 200 ohms. If R_D is significantly greater than this, the N-channel resistance, r_{DS}, is sometimes assumed to be negligible.

(a) What switch condition would this represent, and what will be the drain to source voltage (V_{DS})?

_____

(b) As the gate becomes more negative with respect to the source, the resistance of the N-channel increases until the cutoff point is reached. At this point the resistance of the channel is assumed to be infinite. What condition will this represent, and what will be the drain to source voltage?

_____

(c) What does the JFET act like when it is operated between the two extremes of current saturation and current cutoff?

_____

— — — — — — — — — — — — — — — —

(a) closed switch, $V_{DS} = 0$ V, or very low value
(b) open switch, $V_{DS} = V_{DD}$
(c) a variable resistance

SUMMARY

At this point it may useful to compare the properties of a mechanical switch with those of both types of transistors.

SWITCH	BJT	JFET
OFF or open. No current. Full voltage across terminals.	No collector current. Full supply voltage from collector to emitter.	No drain current. Full supply voltage from drain to source.
ON or closed. Full current. No voltage across terminals.	Full circuit current. Collector to emitter voltage is 0 V.	Full circuit current. Drain to source voltage is 0 V.

The terms ON and OFF are used in digital electronics to describe the two transistor conditions just encountered. Their similarity to a mechanical switch is utilized in many digital situations. In the next chapter we will examine the transistor switch in more detail. This is the first step toward an understanding of digital electronics. In Chapter 8 we will examine the operation of the transistor when it is being biased at a point falling between the two conditions described as ON and OFF. In this mode it can be viewed as a variable resistance and utilized as an amplifier.

SELF-TEST

The questions below will test your understanding of this chapter. Use a separate sheet of paper for your diagrams or calculations. Compare your answers with the answers provided following the test.

1. Draw the symbols for an NPN and a PNP transistor. Label the terminals of each.

2. In the figure below, draw the paths taken by the base and collector currents.

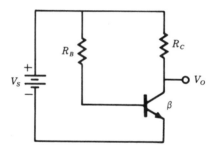

3. What causes the collector current to flow? _____

4. What is meant by the term *current gain?* What symbol is used for this? What is its algebraic formula? _____

_____

5. In the figure in question 2, $R_B = 27$ kΩ and $V_S = 3$ V. Find I_B. Assume that the transistor is made of silicon.

_____

6. Repeat question 5 with $R_B = 220$ kΩ and $V_S = 10$ V. Find I_B.

_____

7. Using the same figure, find V_O when $R_B = 100$ kΩ, $V_S = 10$ V, $R_C = 1$ kΩ, and $\beta = 50$.

_____

8. Repeat question 7 with these values: $R_B = 200$ kΩ, $V_S = 10$ V, $R_C = 1$ kΩ, and $\beta = 50$.

9. Now use these values: $R_B = 47$ kΩ, $V_S = 10$ V, $R_C = 500$ ohms, and $\beta = 65$.

10. Now use these values: $R_B = 68$ kΩ, $V_S = 10$ V, $R_C = 820$ ohms, and $\beta = 75$.

---

11. Draw the symbols for the two types of junction field effect transistors and identify the terminals.

12. In the JFET, what controls the flow of current and what is this similar to for a BJT?

---

13. In the JFET common source circuit shown, put in the correct polarities of the power supplies and draw the current path taken by the drain current.

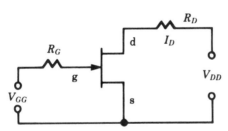

14. Why have we not considered a gate current for the JFET when we had to consider a base current for a BJT?

---

15. Answer the following questions for the circuit shown.

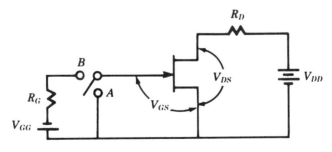

(a) If the switch is at position A, what will the drain current be, and why? _____

(b) If the switch is at position B, and the gate supply voltage is of sufficient value to cause cutoff, what will the drain current be, and why? _____

(c) What is the voltage from the drain to the source for the two switch positions? _____

Answers to Self-Test

If your answers do not agree with those given below, review the frames indicated in parentheses before you go on to the next chapter.

1.

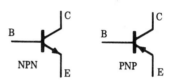

NPN   PNP

(frames 4, 5)

2.

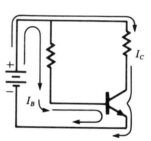

(frames 13, 15)

3. Base current.   (frame 15)

4. Current gain is the ratio of collector current to base current. It is given the symbol β. $\beta = I_C/I_B$.   (frames 16, 17)

5. $I_B = \dfrac{(V_S - 0.7)}{R_B} = \dfrac{(3\text{ V} - 0.7\text{ V})}{27\text{ k}\Omega} = \dfrac{2.3\text{ V}}{27\text{ k}\Omega} = 85\ \mu\text{A}$ (frame 7)

6. $I_B = \dfrac{10\text{ V}}{220\text{ k}\Omega} = 45.45\ \mu\text{A}$ (frame 7)

7. 5 V   (frames 21–24)

8. 7.5 V   (frames 21–24)

9. 3.1 V   (frames 21–24)

10. 1 V   (frames 21–24)

11.

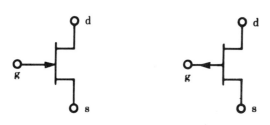

N-channel   P-channel

(frame 30)

12. The voltage on the gate controls the flow of drain current. This is similar to the base voltage controlling the collector current in a BJT. (frame 30)

13.

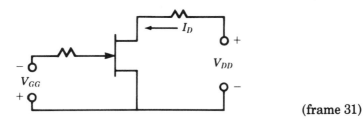

(frame 31)

14. The JFET is a high impedance device and does not draw current from the gate circuit. The BJT is a relatively low impedance device and does, therefore, require some base current to operate it. (frame 29)

15. (a) The drain current will be at its maximum value; in this case it will equal V_{DD}/R_D since we can ignore the drop across the JFET. The gate to source voltage will be 0 V, which reduces the channel resistance to a very small value close to 0 ohms.

(b) The drain current now goes to 0 A because the channel resistance is at infinity (very large), which does not allow electrons to flow through the channel.

(c) At position A, V_{DS} is approximately 0 V. At position B, $V_{DS} = V_{DD}$. (frame 32)

CHAPTER FOUR

The Transistor Switch

In Chapter 3 we saw how the transistor could be turned ON and OFF in a manner similar to a mechanical switch. Almost all industrial controls are now transistor switches, and a computer consists entirely of transistor switches. Computers work with Boolean algebra, which uses only the two logic states— TRUE and FALSE. These two states can easily be represented electronically by a switch which is ON or OFF; thus, the transistor switch is an ideal device for the fast computations and manipulations of Boolean algebra.

Transistors are everywhere; they cannot be ignored in our daily lives. This chapter introduces their most simple and widespread application— switching, with emphasis on the BJT.

When you complete this chapter, you will be able to:

- calculate the base resistance which will turn a transistor ON and OFF;

- explain how one transistor will turn another ON and OFF;

- calculate various currents and resistances in simple transistor switching circuits;

- calculate various resistances and currents in switching circuits which contain two transistors;

- compare the switching action of a JFET to a BJT.

TURNING THE TRANSISTOR ON

1. We will start by examining how to turn a transistor ON, using the simple circuit shown in the following diagram. R_C is a lamp which will be illuminated when the transistor turns ON.

    In the preceding chapter R_B was given, and we had to find the value of collector current and voltages. Now we will do the reverse. We will start with the current through R_C and find the value of R_B which will turn the transistor ON and permit this current to flow.

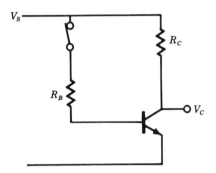

What current values will we have to know in order to find R_B?

_____

— — — — — — — — — — — — — — —

the base and collector currents

2. In the setup in frame 1 the lamp which could be used in place of R_C is referred to as the *load*, and the current through it is called the *load current*.

   (a) Is load current equivalent to base or collector current? _____

   (b) What is the path taken by the collector current in frame 1? Draw it on the circuit.

   — — — — — — — — — — — — — — —

   (a) collector current

   (b)
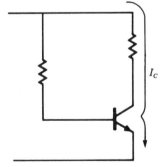

3. For the transistor switch to perform effectively as a CLOSED switch, its collector voltage must be at the same voltage as its emitter. So in this simple circuit the collector voltage will be at ground potential. In this condition the transistor is said to be turned ON.

(a) What is the collector voltage when the transistor is turned ON?

_____

(b) What other component does an ON transistor resemble?

_____

– – – – – – – – – – – – – – – – –

(a) 0 volts
(b) a closed mechanical switch

Note: In actual practice there will be a very small voltage drop across the transistor from the collector to the emitter. This is really a saturation voltage and is the smallest voltage drop that the transistor can have when it is ON as "hard" as possible. In this chapter we are considering it to be a negligible value. Hence, we say the collector voltage is 0 V. For a quality switching transistor this is a good assumption.

4. In the circuit of frame 1, assume a 24 volt lamp is used, and its resistance is 240 ohms.

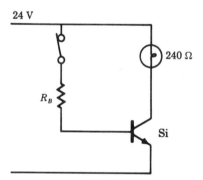

24 V

R_B

240 Ω

Si

We want to calculate the current through the lamp. This is the load or collector current. We have to start with the load because the object is to illuminate the lamp; all the calculations must be aimed to this end. (Also this is the only thing we know.)

$$I_L = I_C = \frac{V_S}{R_C} = \frac{24 \text{ V}}{240 \text{ ohms}} = 100 \text{ mA}$$

We thus require 100 mA of collector current to flow through the transistor so that the lamp will come on. To do this we must provide base current.

(a) Why do we need base current? _____

_____

(b) How will base current be made to flow?

_____

– – – – – – – – – – – – – – – – –

(a) to enable collector current to flow, so that the lamp will light up
(b) by closing the mechanical switch in the base circuit

5. You can calculate the amount of base current flowing. At this point we will assume that $\beta = 100$. (Normally this would be looked up in the manufacturer's data sheets.)
    What is the value of the base current I_B?

_____

$$I_B = \frac{I_C}{\beta} = \frac{100 \text{ mA}}{100} = 1 \text{ mA}$$

6. The base current will flow as shown in the diagram below. When base current is flowing the base-to-emitter terminals behave just as if they were a diode. (In actual fact, they are a diode.)

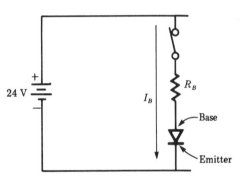

(a) What is the voltage drop across the base-emitter diode? _____
(b) What is the voltage drop across R_B? _____
_____

(a) 0.7 V, since it is a silicon transistor
(b) 24 V if the 0.7 is ignored. 23.3 V if it is not.

7. The next step is to calculate R_B. The current flowing through R_B is the base current I_B, and the voltage across it was determined in frame 6.

    Calculate R_B. _____
_____

$$R_B = \frac{23.3 \text{ V}}{1 \text{ mA}} = 23,300 \text{ ohms}$$

The final circuit with the values included is shown in the figure on page 82.

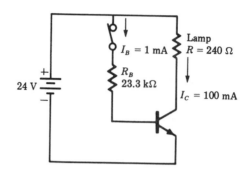

8. The following steps are used in calculating how to turn a transistor ON:

    (1) Determine the collector current.
    (2) Check the value of β.
    (3) Calculate the value of I_B from the results of steps 1 and 2.
    (4) Calculate the value of R_B.
    (5) Draw the final circuit.

    Now assume a 28 V lamp which will draw 50 mA of current will be used. This is typical of many small pilot lamps used in electronic equipment. Assume $\beta = 75$.

    (a) Calculate I_B. _____

    (b) Determine R_B. _____

    — — — — — — — — — — — — — — — — — —

    (a) The collector current and β were given. Thus:

    $$I_B = \frac{I_C}{\beta} = \frac{50 \text{ mA}}{75} = 0.667 \text{ mA}$$

    (b) $R_B = \dfrac{28 \text{ V}}{0.667 \text{ mA}} = 42 \text{ k}\Omega$

    Ignore V_{BE}.

9. Determine R_B when a 9 V lamp is used which draws 20 mA. Assume $\beta = 75$.

    — — — — — — — — — — — — — — — — — —

    $R_B = 31.1 \text{ k}\Omega$
    Now include V_{BE}.

10. You may recall that, in practice, if the supply voltage is much higher than the 0.7 V drop between the base and the emitter, we may ignore the 0.7 V drop and assume that all the supply voltage appears across the base resistor R_B. This simplifies the arithmetic. (Resistors are only accurate to within 10% of their stated value anyway.) If the supply voltage is less than 10 volts, the 0.7 V should not be ignored.

Calculate R_B for the problems following, ignoring the base-emitter drop if appropriate.

(a)   A 10 V lamp which draws 10 mA. $\beta = 100$.

_____

(b)   A 5 V lamp which draws 100 mA. $\beta = 50$.

_____

— — — — — — — — — — — — — — — —

(a) $I_B = \dfrac{10 \text{ mA}}{100} = 0.1 \text{ mA}$

$R_B = \dfrac{10 \text{ V}}{0.1 \text{ mA}} = 100 \text{ k}\Omega$

(b) $I_B = \dfrac{100 \text{ mA}}{50} = 2 \text{ mA}$

$R_B = \dfrac{(5 \text{ V} - 0.7 \text{ V})}{2 \text{ mA}} = \dfrac{4.3 \text{ V}}{2 \text{ mA}} = 2.15 \text{ k}\Omega$

TURNING THE TRANSISTOR OFF

11. Up to now we have concentrated on turning the transistor ON, or making it act like a closed mechanical switch. We will now concentrate on turning it OFF, or making it act like an open mechanical switch. If the transistor is OFF, or OPEN, no current flows through the load, or no collector current flows.

(a)   When a switch is open, are the two terminals at different voltages

or at the same voltage? _____

(b)   When a switch is open does current flow? _____
(c)   For a transistor to turn OFF and act like an open switch, how much

base current is needed? _____

— — — — — — — — — — — — — — — —

(a)   at different voltages, usually the supply voltage and ground voltage, unless another resistor is across the switch
(b)   no
(c)   The transistor is OFF when there is no base current.

Note: When the transistor is OFF as "hard" as possible and no current is said to flow, there is actually a very small leakage current flowing through the transistor. In practice this is often ignored since it is so low, especially in quality switching transistors. This leakage current is, however, specified by manufacturers for different type transistors, as is the small voltage drop that occurs when the transistor is in the ON state.

12. It is easy to ensure that there is no base current. The mechanical switch in the base circuit is just opened, as in the figure below.

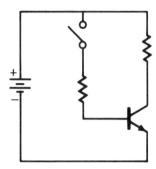

   Opening the mechanical switch will work fairly well, but it does introduce some problems which will not be covered at this time. In practice a slight modification is made to the circuit. A resistor R_2 is placed as shown in the diagram below. This ties the base of the transistor to ground or 0 V, and thus no base current can possibly flow.

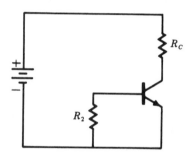

(a) Why will current not flow from the supply voltage to the base-emitter juction? _____

_____

(b) Why will current not flow from collector to base through R_2 to ground? _____

_____

(c) How much current flows from collector to base? _____

(d) Why is the transistor base at 0 volts when R_2 is installed?

_____

– – – – – – – – – – – – – – – –

(a) There is no current path from the supply voltage through the base-emitter junction, thus, there is no base current flowing.
(b) The internal construction of the transistor prevents this.
(c) none at all
(d) because there is no current through R_2

13. Since no current is flowing through R_2 it can be any size between 0
ohms (a short circuit) and infinity (an open circuit). Both of these value
extremes have disadvantages, so some practical value must be chosen.
In practice the values found for R_2 will be between 1 kΩ and 1 MΩ.
  Which of the following values would be practical to keep a transistor turned off? 1 ohm, 2 kΩ, 10 kΩ, 20 kΩ, 50 kΩ, 100 kΩ, 250 kΩ,

500 kΩ. _____

They would all be suitable except the 1 ohm, since the rest are all above
1 kΩ and below 1 MΩ.

14. The figure below shows how to set up the circuit so that the mechanical
switch can be used to turn the transistor ON and OFF. Note that a two-
position switch has now been used.

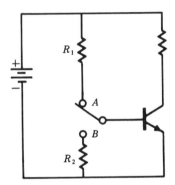

(a) As shown in the diagram, is the transistor ON or OFF? _____

(b) Which position, A or B, will cause the collector current to be

     0 amperes? _____

(a) ON, since base current can exist
(b) position B, since no base current can flow, turning the transistor OFF

  We have used a transistor as a switch instead of using a mechanical
switch to turn a lamp on and off. To turn the transistor ON and make it
behave like a closed mechanical switch we had to make collector current
flow. We did this by providing base current, and calculating the value of R_B
needed.
  To turn the transistor OFF, the base was tied to 0 V through a resistor
R_2, thus ensuring that no collector current would flow, and thus making the
transistor appear to behave like an open switch.

WHY TRANSISTORS ARE USED AS SWITCHES

15. Using the transistor as a switch, as shown in the previous frames, to actually turn a lamp current on and off is one of the simplest and most easily understood uses. Although transistors are often interposed between a mechanical switch and a lamp, this is not the main use for the transistor. The reason this example was chosen was simply to illustrate how easy it is to use a transistor as a switch and how easy the calculations are.

Usually some other item or component in the base circuit does the switching, and usually it is some other item in the collector circuit which is being controlled. Let's look at a few examples that demonstrate the advantages gained by introducing the transistor into a circuit as a switching element.

Example 1. Suppose a lamp is placed in a dangerous area, such as a radioactive chamber. Then obviously the switch must be placed somewhere safe. The transistor provides extra isolation between the chamber and the operator.

Example 2. If a high intensity lamp is used which requires much current, then this current must flow through the wires between the switch and the lamp. By using the transistor as the actual lamp switch and controlling the base current with a smaller switch in the base circuit, only the smaller base current need flow through the wires. This is a much safer arrangement for both the operator and the wiring.

Example 3. A major problem with switching high current in wires is that they produce interference in adjacent wires. This can be disastrous in communications or computer interconnections. Thus, reducing the current to a smaller value is most desirable.

Example 4. Most often it is not a lamp which is switched on and off, as in the examples up to now. It could be some piece of electronic control or communications equipment, or the signal from such equipment. In this case it would be most undesirable to have the signal routed to unnecessary places outside the equipment. It is much easier to use an outside switch to control the signal without the signal ever leaving the equipment. For this purpose transistor switching is ideal.

What features mentioned in the examples make it desirable to use

transistors as switches? _____

_____

— — — — — — — — — — — — — —

It is important to emphasize that the switching action of a transistor can be directly controlled by an electrical signal, as well as a mechanical switch in the base circuit. This gives a lot of flexibility to the design and allows for simple electrical control. Of course, other factors, such as safety, reduction of interference, remote switching control, and even lower design costs are also important. More reasons are given in the next frame.

16. The following examples of transistor switching focus on some other reasons for using transistors.

Example 5. The ON and OFF times of a transistor can be very accurately controlled, while mechanical devices are not very accurate. This is most important in applications such as photography where it is necessary to expose a film or illuminate an object for a precise period of time. In this respect transistors are much more accurate and controllable than any other device known.

Example 6. A transistor can be switched ON and OFF many millions of times a second, and can continue being treated like this almost indefinitely. Under normal operating conditions the transistor is virtually indestructible. Transistors are probably the longest lasting and most reliable component ever known, while mechanical switches usually fail after a few thousand operations.

Example 7. Much industrial control and communications information is now encountered in the form of digital codes. Most of the codes rely upon the presence or absence of a voltage rather than the level of the voltage, and are thus ideally suited to control by switches. By using transistors as switches, very fast and very complex codes are possible.

Example 8. An interesting use is in the medical field. The heart pacemaker has a transistor which causes it to pulse regularly over a long period of time, thus keeping the wearer alive. No mechanical device could possibly do this.

Example 9. Modern manufacturing techniques allow for the miniaturization of transistors to such a great extent that many of them, even thousands, can be fabricated into a single device. These devices are extremely small and are called integrated circuits (ICs). They form the basis for the operation of many of our new products and are applied practically everywhere. Home computers are made possible because of this ability to miniaturize transistor switching circuits.

What other features, besides the ones mentioned in the previous frame, are demonstrated in the examples above? _____

_____

_____

— — — — — — — — — — — — — — —

can be accurately controlled; high speed operation; reliable operation; long life; very small; low power consumption; can be manufactured with large numbers in an extremely small space

17. At this point consider the idea of using one transistor to turn another one ON and OFF, and using the second one to operate the lamp or other load. (This idea is explored in the next section of the chapter.)

If many high current loads are to be switched, then all the switching can be performed with low current switches, and many items can be turned on and off simultaneously with only one remote switch.

(a) With the extra switches added, will the current that flows through the main switch be greater or less than that through the load?

_____

(b) What effect do you think the extra transistor will have on the following?

(1) Safety. _____

(2) Cost of components. _____

(3) Convenience to the operator. _____

_____

(4) Efficiency and smoothness of operation. _____

_____

— — — — — — — — — — — — —

(a) Less current will flow.
(b) (1) It increases safety and provides higher isolation.
     (2) It lowers component cost since low current switches cost less.
     (3) Switches can be placed conveniently close together on a panel, or in the best place for an operator, rather than the switch position dictating operator position.
     (4) One switch can start many things, as in a master lighting panel in a TV studio or theater.

18. Indicate which of the following are good reasons for using a transistor as a switch.

_____ (a) To allow equipment in a dangerous or inaccessible area to be switched on and off.

_____ (b) To switch very low currents or voltages.

_____ (c) To lessen the electrical noise which might be introduced into communication and other circuits.

_____ (d) To increase the number of control switches.

_____ (e) To use a faster, more reliable device than a mechanical switch.

— — — — — — — — — — — — —

a, c, and e.

THE TWO TRANSISTOR SWITCHING CIRCUIT

19. All modern electronic equipment consists of multiple switching transistors, in which one transistor is used to switch others ON and OFF. To illustrate how this works we will continue with the lamp as the load and the mechanical switch as the actuating element. The figure below shows a simple circuit which we will analyze.

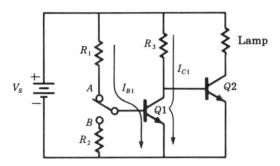

The following description of how this circuit works is typical of that found in many service manuals for professional electronic equipment.

If the switch is in position A, then base current I_{B1} flows through R_1 into the base of $Q1$ and turns it ON. This causes collector current to I_{C1} to flow, and the collector voltage drops to 0 V. Since the base of $Q2$ is connected to the collector of $Q1$, the base of $Q2$ also drops to 0 volts. This ensures that $Q2$ is cut OFF. All of the current through R_3 flows as collector current I_{C1} to ground through $Q1$, thus there is no base current I_{B2} for $Q2$, so it cannot turn ON. Hence the lamp remains dark.

Now assume the switch is in position B, as shown below. No base current can flow into $Q1$, thus no collector current can flow, and so $Q1$ is OFF. In this case the current which flows through R_3 is all base current I_{B2} for $Q2$, which turns $Q2$ ON. This illuminates the lamp.

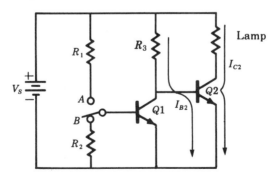

Now that you have read the descriptions of how the circuit works, try to answer the following questions. First assume the switch is in position A, as shown in the first diagram in this frame.

(a) What effect does I_{B1} have on transistor $Q1$? _____

_____

(b) What effect in turn does this have on:

    (1) collector current I_{C1}? _____

    (2) collector voltage V_{C1}? _____

(c) And what effect does this have on:

    (1) the base voltage of $Q2$? _____

    (2) transistor $Q2$ (i.e., is it ON or OFF)? _____

(d) Where does the current through R_3 go? _____

(e) Is the lamp on or off? _____

– – – – – – – – – – – – – –

(a) It turns $Q1$ ON.
(b) (1) I_{C1} flows; (2) V_{C1} drops to 0 V
(c) (1) base of $Q2$ drops to 0 V; (2) $Q2$ is OFF
(d) I_{C1} flows through $Q1$ to ground
(e) off

20. Now assume the switch is in the B position, and try to answer these questions.

(a) How much base current I_{B1} flows into $Q1$? _____

(b) Is $Q1$ ON or OFF? _____

(c) What current flows through R_3? _____

(d) Is $Q2$ ON or OFF? _____

(e) Is the lamp on or off? _____

– – – – – – – – – – – – – – – –

(a) none
(b) OFF
(c) I_{B2}
(d) ON
(e) on

21. Refer back to the two diagrams in frame 19. Now try to answer these questions. Let the supply voltage be 10 V.

(a) Is the current through R_3 ever divided between $Q1$ and $Q2$?

    Explain. _____

_____

(b) What is the collector voltage of $Q2$ with the switch in each position? _____

(c) What is the collector voltage of $Q1$ with the switch in each position? _____

- - - - - - - - - - - - - - - -

(a) No. If $Q1$ is ON all the current flows through it to ground as collector current. If $Q1$ is OFF all the current flows through the base of $Q2$ as base current.

(b) In position A: 10 volts since it is OFF
In position B: 0 volts since it is ON

(c) In position A: 0 volts since it is ON
In position B: The collector voltage of $Q1$ will equal the voltage drop across the forward biased base-emitter junction of $Q2$, since it is in parallel with it. It will not rise to 10 V, but can only rise to 0.7 V if $Q2$ is made of silicon.

22. We will now calculate the values of R_1, R_2, and R_3 for this circuit. The process is similar to the one used before, but is expanded to deal with the extra transistor. Note that we will still begin with the load current. The calculation steps are outlined as follows.

(1) Determine the load current I_{C2}.
(2) Determine β for $Q2$. Call this β_2.
(3) Calculate I_{B2} for $Q2$. Use $I_{B2} = I_{C2}/\beta_2$.
(4) Calculate R_3 to provide this base current. Use $R_3 = V_S/I_{B2}$.
(5) R_3 is also the load for $Q1$ when $Q1$ is ON.
Therefore, the load current for $Q1$ will have the same value as the base current for $Q2$, as calculated in step 3.
(6) Determine β_1, the β for $Q1$.
(7) Calculate the base current for $Q1$. Use $I_{B1} = I_{C1}/\beta_1$.
(8) Find R_1. Use $R_1 = V_S/I_{B1}$
(9) Choose R_2. For convenience, let $R_2 = R_1$.

We will continue to work with the same circuit. It is shown on the next page with the following values.

A 10 volt lamp which draws 1 ampere
$\beta_2 = 20$, $\beta_1 = 100$

Ignore any voltage drops across the transistors. Calculate the following.

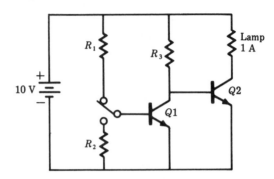

(a) Find I_{B2} as in step 3.

$I_{B2} =$ _____

(b) Find R_3 as in step 4.

$R_3 =$ _____

(c) Calculate the load current for Q1 when it is ON as shown in step 5.

$I_{C1} =$ _____

(d) Find the base current for Q1.

$I_{B1} =$ _____

(e) Find R_1 as in step 8.

$R_1 =$ _____

(f) Choose a suitable value for R_2.

$R_2 =$ _____

– – – – – – – – – – – – – – – – – –

The answers here correspond to all the steps.

(a) (1) The load current is given as 1 A.
    (2) $\beta_2 = 20$ (given). This is a typical value for a transistor which would handle 1 A.

    (3) $I_{B2} = \dfrac{1\,A}{20} = 50\,mA$

(b) (4) $R_3 = \dfrac{10\ volts}{50\ mA} = 200\ \Omega$

    Note the 0.7 V base-emitter drop has been ignored.
(c) (5) $I_{C1} = I_{B2} = 50\,mA$
(d) (6) $\beta_1 = 100$ (again a typical value)

    (7) $I_{B1} = \dfrac{50\ mA}{100} = 0.5\,mA$

(e) (8) $R_1 = \dfrac{10 \text{ V}}{0.5 \text{ mA}} = 20 \text{ k}\Omega$

Again the 0.7 V drop is ignored.

(f) (9) For convenience choose R_2 to be the same as R_1, or 20 kΩ. This will lessen the number of different components in the circuit. Fewer different components makes manufacturing and inventory control easier. You could of course choose any value between 1 kΩ and 1 MΩ.

23. Following the same procedure, with the same circuit, work through another example. Use a 28 volt, 560 mA lamp. Assume $\beta_2 = 10$ and $\beta_1 = 100$. Calculate the following.

(a) I_{B2} ————————————————

(b) R_3 ————————————————

(c) I_{C1} ————————————————

(d) I_{B1} ————————————————

(e) R_1 ————————————————

(f) R_2 ————————————————

— — — — — — — — — — — — — — — —

(a) 56 mA
(b) 500 ohms
(c) 56 mA
(d) 0.56 mA
(e) 50 kΩ
(f) 50 kΩ by choice

THE THREE TRANSISTOR SWITCH

24. To include a third transistor is an extension of the two transistor switching circuit. The diagram on the next page shows such a circuit.
Q1 is used to turn Q2 ON and OFF, and Q2 is used to operate Q3. All of the circuit descriptions and calculations are as before, but one extra stage of work is needed to deal with the third transistor. Look at the circuit and answer the following questions.

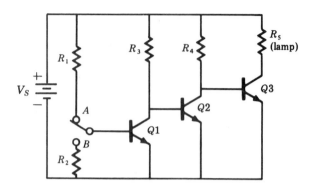

(a)   If the switch is in position A, is $Q1$ ON or OFF? _____

(b)   In the position A, is $Q2$ ON or OFF? _____

(c)   Where is current through R_4 flowing? _____

(d)   Is Q_3 ON or OFF? _____

_ _ _ _ _ _ _ _ _ _ _ _ _ _ _ _ _

(a)   ON
(b)   OFF
(c)   Into the base of $Q3$
(d)   ON

25.   If the switch is in position B, $Q1$ is held OFF.

(a)   Is $Q1$ ON or OFF? _____

(b)   Is $Q2$ ON or OFF? _____

(c)   Where is the current through R_4 flowing? _____

(d)   Is $Q3$ ON or OFF? _____

(e)   Which switch position will turn the lamp on? _____

(f)   How does the switch position in the three transistor switch differ

from that in the two transistor switch circuit? _____

_____

_ _ _ _ _ _ _ _ _ _ _ _ _ _ _ _ _

(a)   OFF
(b)   ON
(c)   through $Q2$ to ground
(d)   OFF
(e)   position A
(f)   The positions are opposite. The addition of an extra transistor to
      change the operating state of a switch is quite common in
      electronics.

26. We will work through an example with values, just as before. The steps are similar but a few have been added, as shown below.

(1)   Find the load current. This is often given.
(2)   Determine the current gain of $Q3$. This is β_3 and usually it is a given value.
(3)   Calculate I_{B3}.
(4)   Calculate R_4.
(5)   Assume $I_{C2} = I_{B3}$.
(6)   Find β_2. Again this is a given value.
(7)   Calculate I_{B2}.
(8)   Calculate R_3.
(9)   Assume $I_{C1} = I_{B2}$.
(10) Find β_1.
(11) Calculate I_{B1}.
(12) Calculate R_1.
(13) Choose R_2.

For this example we will use a 10 volt lamp which draws 10 amperes. Assume the βs of the transistors are given in the manufacturer's data sheets as $\beta_1 = 100$, $\beta_2 = 50$, and $\beta_3 = 20$. Now work through the steps, checking the answers for each step as you complete it.

– – – – – – – – – – – – – – – –

(1)   The load current is given as 10 A.
(2) β_3 is given as 20.

(3) $I_{B3} = \dfrac{I_{C3}}{B_3} = \dfrac{10 \text{ A}}{20} = 0.5 \text{ A} = 500 \text{ mA}$

(4) $R_4 = \dfrac{10 \text{ volts}}{500 \text{ mA}} = 20 \text{ ohms}$

(5) $I_{C2} = I_{B3} = 500 \text{ mA}$
(6) β_2 is given as 50.

(7) $I_{B2} = \dfrac{I_{C2}}{\beta_2} = \dfrac{500 \text{ mA}}{50} = 10 \text{ mA}$

(8) $R_3 = \dfrac{10 \text{ V}}{10 \text{ mA}} = 1 \text{ k}\Omega$

(9) $I_{C1} = I_{B2} = 10 \text{ mA}$
(10) β_1 is given as 100.

(11) $I_{B1} = \dfrac{I_{C1}}{\beta_1} = \dfrac{10 \text{ mA}}{100} = 0.1 \text{ mA}$

(12) $R_1 = \dfrac{10 \text{ V}}{0.1 \text{ mA}} = 100 \text{ k}\Omega$

(13) R_2 can be chosen to be 100 kΩ also.

27. Determine the values in the circuit for a 75 volt lamp which draws 6 A. Assume $\beta_3 = 30$, $\beta_2 = 100$, and $\beta_1 = 120$. Calculate the following.

(a) I_{B3} _____

(b) R_4 _____

(c) I_{B2} _____

(d) R_3 _____

(e) I_{B1} _____

(f) R_1 _____

(g) R_2 _____

– – – – – – – – – – – – – – – –

If you followed the steps in frame 26, you should have these values.

(a) 200 mA
(b) 375 Ω
(c) 2 mA
(d) 37.5 kΩ
(e) 16.7 μA
(f) 4.5 MΩ
(g) Choose $R_2 = 1$ MΩ

ALTERNATIVE BASE SWITCHING

28. In the examples of transistor switching, we saw the actual switching performed with a small mechanical switch placed in the base circuit of the first transistor. This switch has been shown as having three terminals, and as switching from position A to position B. (It is a single pole double throw switch.) It does not have a definite ON or OFF position like a simple ON-OFF switch.

    Why couldn't a simple ON-OFF switch with only two terminals have been used with those examples? _____

    _____

– – – – – – – – – – – – – – – –

It could not change over from one position to the other.

29. In order to use a simpler switch in a transistor switching circuit, the circuit must be modified as shown on the next page.

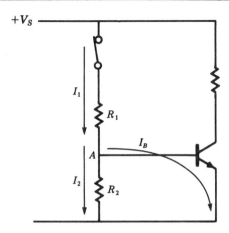

(a) When the switch is open, is Q1 ON or OFF? _____

(b) When the switch is closed, is the lamp on (lit) or off? _____

— — — — — — — — — — — — — — — —

(a) OFF
(b) on

30. When the switch is closed, current flows through R_1. However, at point A in the diagram above the current divides into two paths. One path is the base current I_B, and the other is marked I_2.

   How could you calculate the total current I_1? _____

— — — — — — — — — — — — — — — —

$I_1 = I_B + I_2$

31. The problem now is to choose the values of both R_1 and R_2 so that when the current divides there will be sufficient base current to turn Q1 ON.
   Consider this simple example. Assume the load is a 10 volt lamp which needs 100 mA of current, and $\beta = 100$. Find the base current needed.

$I_B =$ _____

— — — — — — — — — — — — — —

$I_B = \dfrac{100 \text{ mA}}{100} = 1 \text{ mA}$

32. When the current I_1 flows through R_1 it must divide, and 1 mA of it becomes I_B. The remainder will be I_2. The difficulty at this point is that there is no unique value for either I_1 or I_2. In other words, they could

be almost any values. The only restriction is that both must permit 1 mA to flow into the base of $Q1$.

An arbitrary choice must be made for these two values. Based on practical experience, it is common to set I_2 to be 10 times greater than I_B. This split makes the circuits work reliably and still keeps the calculations easy. Thus:

$$I_2 = 10I_B$$
$$I_1 = 11I_B$$

In frame 31 we determined that $I_B = 1$ mA. So what is the value of I_2?

_____

– – – – – – – – – – – – – – – –

$I_2 = 10$ mA

33. Now it is possible to calculate the value of R_2. The voltage across R_2 will be the same as the voltage between the base and the emitter of $Q1$. Assume we are using a silicon transistor, so this is 0.7 V.

(a)   What is the value of R_2? _____

(b)   What is the value of R_1? _____

– – – – – – – – – – – – – – – –

(a) $R_2 = \dfrac{0.7 \text{ V}}{10 \text{ mA}} = 70$ ohms

(b) $R_1 = \dfrac{(10 \text{ V} - 0.7 \text{ V})}{11 \text{ mA}} = \dfrac{9.3 \text{ V}}{11 \text{ mA}} = 800$ ohms (approximately)

The 0.7 V could be ignored here; this would give $R_1 = 910$ ohms.

34. The values you calculated in frame 33 will ensure that the transistor turns ON and illuminates the lamp. The labeled circuit is shown below.

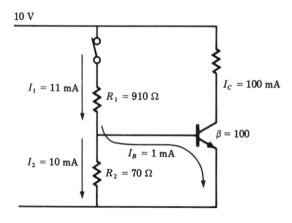

For each of the lamps below, perform the same calculations to find the values of R_1 and R_2.

(a)   28 volt lamp which draws 56 mA. $\beta = 100$

_____

_____

(b)   12 volt lamp which draws 140 mA. $\beta = 50$

_____

_____

— — — — — — — — — — — — — — — — — —

(a) $I_B = \dfrac{56 \text{ mA}}{100} = 0.56 \text{ mA}$

$I_2 = 5.6 \text{ mA}$

$R_2 = \dfrac{0.7 \text{ V}}{5.6 \text{ mA}} = 125 \text{ ohms}$

$R_1 = \dfrac{28 \text{ V}}{6.16 \text{ mA}} = 4.5 \text{ k}\Omega$

(b) $R_2 = 25 \text{ ohms}$
$R_1 = 400 \text{ ohms}$

35.   The arbitrary decision to place a value on I_2 of 10 times the value of I_B is obviously subject to considerable discussion, doubt, and disagreement. *Transistors are not exact devices;* they are not carbon copies of each other. In general, any one of the same type will have a different β from any other. This leads to a certain inexactness in designing and analyzing transistor circuits. If exact mathematical procedures from a textbook are followed, a great deal of computation must be performed, and there will inevitably come a point where an assumption about some figure will have to be made. In addition, the exact figures produced in textbook-type examples have little relation to reality due to the tolerances found in component manufacture.

In practice, then, a few "rules of thumb" have been developed to make the necessary assumptions; these rules lead to simple equations which provide values for components which will work in practice.

It makes life easier to use these rules of thumb. We have used one simple such rule here. It should be understood that these rules of thumb have not been chosen capriciously. They will all stand up to rigorous mathematical analysis.

Is the choice of $I_2 = 10I_B$ the only choice which will work? Most emphatically not. Almost any value of I_2 which is at least 5 times larger than I_B will give satisfactory results. Choosing 10 times the value is good for three reasons.

(1) It is a good practical choice—it will always work.
(2) It makes arithmetic easy.
(3) It does not complicate the understanding of the circuits by encumbering the new student with unnecessary mathematics.

In the example from frame 32, $I_B = 1$ mA and we used 10 mA for I_2. Which of the following values would also have worked?

____ (a) 5 mA   ____ (d) 6.738 mA

____ (b) 8 mA   ____ (e) 1 mA

____ (c) 175 mA

- - - - - - - - - - - - - - - - - -

Choices a, b, and d are good. Value c is much too high to be a sensible choice, and e is too low.

36. Before we finish this chapter, answer the following review questions.

(a) Which switches faster, the transistor or the mechanical switch?

_____

(b) Which can be more accurately controlled? _____

(c) Which is the easiest to operate remotely? _____

(d) Which is the most reliable? _____

(e) Which has the longest life? _____

- - - - - - - - - - - - - - - - - -

(a) The transistor is much faster.
(b) The transistor.
(c) The transistor.
(d) The transistor.
(e) Transistors are indestructible under normal operating conditions, and they do not wear out since they have no moving parts. A mechanical switch will fail after several thousand operations. Transistors will operate several million times a second and continue to do so for years.

SWITCHING THE JFET

37. The JFET (junction field effect transistor), which was introduced in Chapter 3, can now be compared with the BJT (bipolar junction transistor). You may want to review frames 29 through 32 of Chapter 3 which introduced this device. The JFET is considered a "normally on" device, which means that with no voltage applied to the input terminal (called the gate) it is in a highly conductive state, offering low resistance to current flow. When a voltage is applied to the gate, the device

conducts less current because the resistance of the drain to source channel has increased. At some point, as the voltage increases, the resistance can become so large that the device will "cut off" the flow of current.

(a)   What are the three terminals for a JFET called, and which one controls the operation of the device? _____

_____

(b)   What determines the ON and OFF nature of the JFET? _____

_____

— — — — — — — — — — — — — — — —

(a)   drain, source, and gate, with the gate acting as the control
(b)   When the gate voltage is zero (at the same potential as the source), the drain current is at its highest value (device ON). When the gate to source voltage is high, the drain current is 0 A (device OFF).

38.   The circuit shown here will explore the switching ability of the JFET. Note that there is a variable resistor on the gate side of the JFET. If you can obtain a JFET and some equipment, you should be able to follow this discussion experimentally and relate the operation of the JFET to the characteristic curve shown.

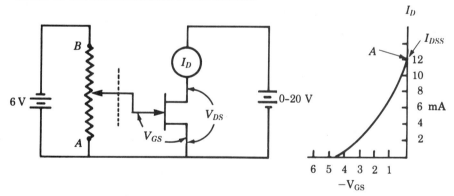

With the gate resistor set at point A, the voltage from the gate to the source is zero ($V_{GS} = 0$). The current that flows in the drain at this time is called the saturation current (I_{DSS}). We can adjust V_{DD} over a range of values and see that the value of I_{DSS} stays constant. At some low value of V_{DS} we will see the current begin to reduce in value. This point is called the pinch-off point. We will operate the JFET at a level above this pinch-off value. Set the drain supply at about 12 V, to insure that we are in the operating region. On the graph shown, we will be at the point marked with an A. This graph is called the transfer, or transconductance, curve and is a characteristic curve that is supplied by the manufacturer for each type JFET. Now, slowly adjust the gate resistor toward B. This will apply a negative voltage to the gate with respect to the source. We will see a reduction in the drain current as the voltage becomes more negative. If we record some data points and plot them, (I_D vs. V_{GS}), we will see a curve similar to the typical curve shown. At some voltage value, the drain current will be at 0 A. On the graph this is at the point where the curve reaches the V_{GS} axis. Using the transfer curve shown, answer the following:

(a)   With $V_{GS} = 0$, what is the value of drain current? _____

(b)   Why is this value called the drain saturation current? _____

_____

(c)   What is the gate to source cutoff voltage for the curve shown? __

_____

(d)   Why is this called a cutoff voltage? _____

_____

(a)   12 mA on the graph
(b)   The word saturation is used to indicate that the current will remain at this value as V_{DS} is varied above the pinch-off value.
(c)   approximately -4.2 V on the graph
(d)   It is termed a cutoff voltage because at this value the drain current goes to 0 A.

39.   Now let's look at the following circuit and assume that the JFET has the transfer characteristic shown by the curve in frame 38.

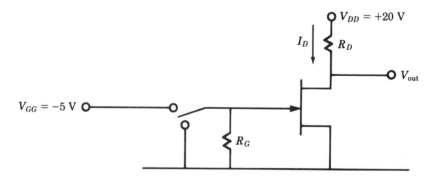

When the gate is connected to ground, the drain current will be at 12 mA (see graph, frame 38). With the assumption that the drain to source resistance is negligible, the required value for R_D can be calculated from

$$R_D = \frac{V_{DD}}{I_{DSS}}$$

If a drain to source voltage is known, then it can be included in the calculation by

$$R_D = \frac{(V_{DD} - V_{DS})}{I_{DSS}}$$

What should the value of R_D be for the characteristic I_{DSS} given by the curve? _____

_ _ _ _ _ _ _ _ _ _ _ _ _ _ _ _

$$R_D = \frac{20 \text{ V}}{12 \text{ mA}} = 1.67 \text{ k}\Omega$$

40. For the JFET circuit of frame 39, let us now assume V_{DS} actually equals 1 V when the N-channel is experiencing current saturation.

   (a) What is the required value of R_D? _____

   (b) What is the effective drain to source resistance (r_{DS}) in this

   situation? _____

_ _ _ _ _ _ _ _ _ _ _ _ _ _ _ _ _

   (a) $R_D = \frac{(20 - 1)}{12 \text{ mA}} = 1583 \text{ ohms}$

   (b) $r_{DS} = \frac{V_{DS}}{I_{DSS}} = \frac{1}{12 \text{ mA}} = 83 \text{ ohms}$

Note: We can see from this calculation that R_D is 19 times greater than r_{DS}. Thus, ignoring V_{DS} and assuming that $r_{DS} = 0$ does not greatly affect the value of R_D. The 1.67 kΩ value is only about 5% higher than the 1583 ohm value for R_D.

41. Now let us turn the JFET OFF. From the graph we can see that a cutoff value of -4.2 V will be required. We will use a gate to source value of -5 V to insure that the JFET will be in the "hard OFF" state. The purpose of resistor R_G is to make sure that the gate is connected to ground during the time of changing the gate voltage from one level to the other. A large value of 1 MΩ is used here to keep from drawing any appreciable current from the gate supply.

   When the gate is at the -5 V potential, what is the drain current

   and the resultant output voltage? _____

_ _ _ _ _ _ _ _ _ _ _ _ _ _ _ _

$I_D = 0$ A and $V_{\text{out}} = V_{DS} = 20$ V, which is V_{DD}

SUMMARY

In this chapter the reader has been introduced to the simplicity of the transistor switch and has learned how to use it and how to calculate the resistor values required for its use.

A lamp was used as the load example, since this gives an easy visual demonstration of the switching action when the circuit is set up on a workbench. If you have the facilities, all of the circuits shown in this chapter will work, and the voltage and current measurements will be very close to those in the text.

You have not yet learned all there is to transistor switching. For example we haven't found out: how much current a transistor will pass before it will burn out; what the maximum voltage is it will stand; or how fast a transistor can switch ON and OFF. These things can be learned from other books; we will not cover them here.

The JFET, when used as a switch, will not switch as fast as a BJT, but will have certain advantages relating to its large input resistance. It does not draw any current from the control circuit in order to operate. Hence, the control circuit always sees a large resistance. The control circuit for the BJT will see a small input resistance when the BJT is in the ON state.

SELF-TEST

The questions below will test your understanding of this chapter. Use a separate sheet of paper for your diagrams or calculations. Compare your answers with the answers provided following the test.

In the first three questions the circuit shown below is used, and the object is to find the value of R_B which will turn the transistor ON. As you may know, resistors are manufactured with "standard values." After an exact value has been calculated, the nearest standard value can be chosen, if you happen to know it, or care to look it up in the appendix.

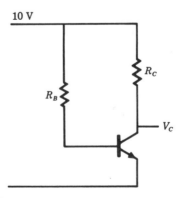

1. $R_C = 1$ kΩ, $\beta = 100$. $R_B =$ _____

2. $R_C = 4.7$ kΩ, $\beta = 50$. $R_B =$ _____

3. $R_C = 22$ kΩ, $\beta = 75$. $R_B =$ _____

For questions 4–6, the circuit shown below is used. Find the values of R_3, R_2, and R_1 which will ensure that $Q2$ is turned ON and OFF. Calculate the resistors in the order given. After the exact values have been found, again choose the nearest standard values.

You should be aware also that rounding off throughout a problem, or just at the final answer, could produce slightly different results.

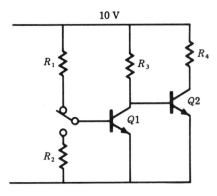

4. $R_4 = 100$ ohms, $\beta_1 = 100$, $\beta_2 = 20$.

   $R_3 =$ _____ $R_1 =$ _____ $R_2 =$ _____

5. $R_4 = 10$ ohms, $\beta_1 = 50$, $\beta_2 = 20$.

   $R_3 =$ _____ $R_1 =$ _____ $R_2 =$ _____

6. $R_4 = 250$ ohms, $\beta_1 = 75$, $\beta_2 = 75$.

   $R_3 =$ _____ $R_1 =$ _____ $R_2 =$ _____

For questions 7–9 find the values of the resistors in the circuit below which will turn $Q3$ ON and OFF. Then select the nearest standard values. As before do the calculations in the reverse numerical order.

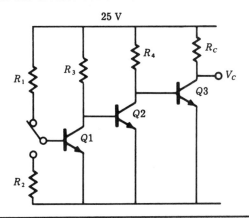

7. $R_C = 10$ ohms, $\beta_3 = 20$, $\beta_2 = 50$, $\beta_1 = 100$.

$R_4 = $ _____ $R_2 = $ _____

$R_3 = $ _____ $R_1 = $ _____

8. $R_C = 28$ ohms, $\beta_3 = 10$, $\beta_2 = 75$, $\beta_1 = 75$.

$R_4 = $ _____ $R_2 = $ _____

$R_3 = $ _____ $R_1 = $ _____

9. $R_C = 1$ ohm, $\beta_3 = 10$, $\beta_2 = 50$, $\beta_1 = 75$.

$R_4 = $ _____ $R_2 = $ _____

$R_3 = $ _____ $R_1 = $ _____

Questions 10–12 use the circuit shown below. Find R_1 and R_2 which will turn the transistor ON and OFF.

10 V

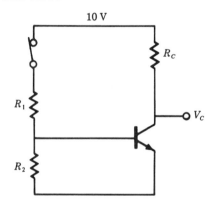

10. $R_C = 1$ kΩ, $\beta = 100$.

$R_1 = $ _____ $R_2 = $ _____

11. $R_C = 22$ kΩ, $\beta = 75$.

$R_1 = $ _____ $R_2 = $ _____

12. $R_C = 100$ Ω, $\beta = 30$.

$R_1 = $ _____ $R_2 = $ _____

13. An N-channel JFET has a transfer curve with the following characteristics. When $V_{GS} = 0$ V, the saturation current, (I_{DSS}) is 10.5 mA. The
cutoff voltage is -3.8 V. With a drain supply of 20 V, design a biasing
circuit that will switch the JFET from the ON state to the OFF state.

Answers to Self-Test

The exercises in this Self-Test show calculations which are typical of those found in electronic practice, and the odd figures sometimes produced are quite common. Thus choosing a nearest standard value of resistor is a common practice. If your answers do not agree with those given below, review the frames indicated in parentheses before you go on to the next chapter.

1. $100 \text{ k}\Omega$ (frame 8)

2. $235 \text{ k}\Omega$. Choose $240 \text{ k}\Omega$ as a standard value. (frame 8)

3. $1.65 \text{ M}\Omega$. Choose $1.6 \text{ M}\Omega$ as a standard value. (frame 8)

4. $R_3 = 2 \text{ k}\Omega$; $R_1 = 200 \text{ k}\Omega$; $R_2 = 200 \text{ k}\Omega$. Use these values.   (frame 22)

5. $R_3 = 200$ ohms; $R_1 = 10 \text{ k}\Omega$; $R_2 = 10 \text{ k}\Omega$. Use these values. (frame 22)

6. $R_3 = 18.8 \text{ k}\Omega = 18 \text{ k}\Omega$ standard value; $R_1 = 1.41 \text{ M}\Omega = 1.5 \text{ M}\Omega$.
    Select $1 \text{ M}\Omega$ for R_2. (frame 22)

7. $R_4 = 200$ ohms; $R_3 = 10 \text{ k}\Omega$; $R_2 = 1 \text{ M}\Omega$; $R_1 = 1 \text{ M}\Omega$.
    Use these values. (frame 26)

8. $R_4 = 280$ ohms $= 270$ ohms standard value; $R_3 = 21 \text{ k}\Omega = 22 \text{ k}\Omega$;
    $R_2 = 1.56 \text{ M}\Omega = 1.5$ or $1.6 \text{ M}\Omega$; $R_1 = 1.56 \text{ M}\Omega = 1.5$ or $1.6 \text{ M}\Omega$
    (frame 26)

9. $R_4 = 10$ ohms $= 10$ ohms standard value; $R_3 = 500$ ohms $= 510$ ohms;
    $R_2 = 37.5 \text{ k}\Omega = 39 \text{ k}\Omega$; $R_1 = 37.5 \text{ k}\Omega = 39 \text{ k}\Omega$. (frame 26)

10. $R_2 = 700$ ohms $= 680$ or 720 ohms standard value; $R_1 = 8.45 \text{ k}\Omega =$
    $8.2 \text{ k}\Omega$. If 0.7 is ignored, then $R_1 = 9.1 \text{ k}\Omega$. (frames 31–33)

11. $R_2 = 11.7 \text{ k}\Omega = 12 \text{ K}$ standard value; $R_1 = 141 \text{ k}\Omega = 140$ or $150 \text{ k}\Omega$.
    (frames 31–33)

12. $R_2 = 21$ ohms $= 22$ ohms standard value; $R_1 = 273$ ohms $= 270$ ohms.
    (frames 31–33)

13. Use the circuit of frame 39. Set the gate supply at a value slightly more negative than -3.8 V. A value of -4 V would work. Make resistor $R_G = 1 \text{ M}\Omega$. Set R_D at a value of $20/10.5$ mA, which calculates a resistance of $1.9 \text{ k}\Omega$. A standard resistor of $1 \text{ k}\Omega$ in series with the standard value of 910 ohms would do nicely. (frames 39 and 41)

CHAPTER FIVE

AC Pre-Test and Review

Some basic knowledge of alternating current (AC) is needed for electronics. The study of AC is basically a study of sine waves and what happens to them. Sine waves are a perfectly natural occurrence and are found in many places. Perhaps the most common is the house current provided at a wall plug. This is a 120 volt sine wave which changes at a rate of 60 cycles per second, or at a frequency of 60 Hz (Hertz). This current is produced in a power station by a large piece of rotating machinery called a generator. A smaller generator is found in a car to provide electrical power for the engine and to charge the battery.

The sound from most musical instruments consists of sine waves. And many other electronic signals—such as speech—are complex combinations of many sine waves, all at different frequencies.

The study of AC starts with the properties of simple sine waves, and continues with variations in sine wave currents caused by differing voltages. In electronics many sine wave voltages and currents are produced by circuits called oscillators, and these signals may change many millions of times a second.

In this chapter we will review:

- the generator;

- the sine wave;

- peak-to-peak and RMS voltages;

- frequency and period;

- resistors in AC circuits;

- capacitive and inductive reactance;

- resonance.

THE GENERATOR

1. In DC work the voltage source is usually a battery, which produces a steady constant voltage and a steady constant current through a conductor.

    In AC work the voltage source is usually a *generator*, which produces a regular output waveform, such as a sine wave.

    Draw one cycle of a sine wave.

_ _ _ _ _ _ _ _ _ _ _ _ _ _ _ _ _

2. There are a number of electronic instruments that are used in the laboratory to produce sine waves. They allow you to adjust the voltage and frequency by turning a dial or pushing a button. These instruments have various names, generally based on their method of producing the sine wave or their application as a test instrument. The most popular generator at present is called a *function generator*. It actually provides a choice of functions or waveforms, including a square wave and a triangle wave. These waveforms are useful in testing certain electronic circuits.

    Modern function generators are a result of the application of much of the modern circuitry that is referred to in this book. For our study, the term "generator" will simply mean a sine wave source. The following symbol is used for a generator. Note that the sine wave shown within the circle designates that it is an AC sine wave source.

(a)   What is the most popular instrument used in the lab to provide waveforms? _____

(b)   When we say AC, what are we referring to? _____

_____

(c)   What does the sine wave inside the generator symbol indicate?

_____

_ _ _ _ _ _ _ _ _ _ _ _ _ _ _

(a)   the function generator
(b)   alternating current, as opposed to direct current
(c)   The generator is a sine wave source.

3.   In the sine wave diagrammed below, certain parameters are marked. The two axes are voltage and time.

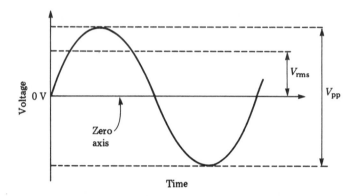

All of the voltage measurements are made from the *zero axis*. This is also known as the *base line* or the *datum line*.

(a)   What is the base line here? _____

(b)   What is the usual point for making time measurements? _____

_____

– – – – – – – – – – – – – – – –

(a)   The base line is the zero axis or datum line.
(b)   Time measurements can be made from any point in the sine wave, but usually they are made from the point at which it crosses the zero axis.

4.   The three most important voltage or amplitude measurements are the *peak* (p), *peak-to-peak* (pp), and the *root mean square* (rms) voltages.
    The formula for the relationship between p, pp, and rms voltages is shown below.

$$V_{pp} = 2V_p = 2 \times \sqrt{2} \times V_{rms}$$

Note: $\sqrt{2} = 1.414$

$$\frac{1}{\sqrt{2}} = 0.707$$

Refer to the waveform diagram in frame 3. If the pp amplitude is 10 volts, find the rms voltage. _____

– – – – – – – – – – – – – – –

$$V_{rms} = \frac{1}{\sqrt{2}} \times \frac{V_{pp}}{2} = 0.707 \times \frac{10}{2} = 3.535 \text{ V}$$

5.   If the rms voltage is 2 volts, find the pp voltage. —————————

———————————————

$V_{pp} = 2 \times \sqrt{2} \times V_{rms} = 2 \times 1.414 \times 2 = 5.656$ V

6.   (a) $V_{pp} = 220$. Find V_{rms}. —————————————
     (b) $V_{rms} = 120$. Find V_{pp}. —————————
———————————————

     (a)   77.77 volts
     (b)   340 volts (This is the common house current supply voltage;
            340 $V_{pp} = 120 V_{rms}$.)

7.   There is one primary time measurement—the duration of the complete
     sine wave. All other time measurements are fractions or occasionally
     multiples of this.

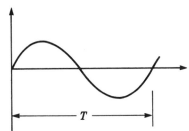

     (a)   What is one complete sine wave called? ————————————
     (b)   What do we call the time to complete one sine wave? ————————
     (c)   How is the frequency related to this time? ————————
     (d)   What is the unit for frequency? ————————————
     (e)   If the period of a sine wave is 0.5 ms, what is its frequency? What
            is the frequency for 40 μsec?

———————————————

     (f)   If the frequency of a sine wave is 60 Hertz, what is its period?
            What about for 12.5 kHz and 1 MHz?

———————————————

     ————————————————

(a)   a cycle
(b)   the period, T
(c) $f = 1/T$
(d)   Hertz (Hz) is now the standard unit for frequency. An older unit that is still sometimes used is cycles per second.
(e)   2 kHz, 25 kHz
(f)   16.7 ms, 80 μsec, 1 μsec

8.   Which of the following could represent electrical AC signals?

—— (a)   A simple sine wave.

—— (b)   A mixture of many sine waves, of different frequencies and amplitudes.

—— (c)   A straight line.

- - - - - - - - - - - - - -

a and b

RESISTORS IN AC CIRCUITS

9.   AC signals are passed through components, just as DC current are. Resistors interact with AC signals just as they do with DC currents.
      Suppose an AC signal of 10 V_{pp} is connected across a 10 ohm resistor. What is the current through the resistor? _____

- - - - - - - - - - - - - -

Use Ohm's law.

$$I = \frac{V}{R} = \frac{10\ V_{pp}}{10\ ohms} = 1\ A\ pp$$

Since the voltage was given in pp, the current is a pp current.

10.   An AC signal of 10 V_{rms} is connected across a 20 ohm resistor. Find the current. _____

- - - - - - - - - - - - - -

$$I = \frac{10\ V_{rms}}{20\ ohms} = 0.5\ A\ rms$$

Since the voltage was given in rms, the current is rms.

11. 10 V_{pp} is applied to the voltage divider circuit shown below. Find V_{out}.

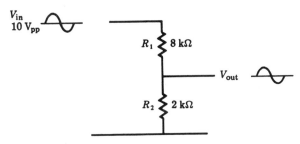

$$V_{out} = V_{in} \times \frac{R_2}{(R_1 + R_2)} = 10 \times \frac{2 \text{ k}\Omega}{(8 \text{ k}\Omega + 2 \text{ k}\Omega)} = 10 \times \frac{2}{10} = 2 \; V_{pp}$$

CAPACITORS IN AC CIRCUITS

12. A capacitor opposes the flow of an AC current.

   (a) What is this opposition to the current flow called? _____

   (b) What is this similar to in DC circuits? _____

   ― ― ― ― ― ― ― ― ― ― ― ― ― ― ―

   (a) reactance
   (b) resistance

13. Reactance is determined by a well known formula.

   (a) What is the formula for reactance? _____

   (b) What does each symbol stand for? _____

   _____

   ― ― ― ― ― ― ― ― ― ― ― ― ― ―

   (a) $X_C = \dfrac{1}{2\pi f C}$

   (b) X_C = the reactance of the capacitor in ohms
       f = the frequency of the signal in hertz.
       C = the value of the capacitor in farads.

14. Assume the capacitance is 1 μF and the frequency is 1 kHz. Find the capacitor's reactance. (Note: $1/2\pi = 0.159$ approximately.)

   _____

   ― ― ― ― ― ― ― ― ― ― ― ― ― ―

$$X_C = \frac{1}{2\pi f C}$$

$f = 1 \text{ kHz} = 10^3 \text{ Hz}$
$C = 1 \mu\text{F} = 10^{-6} \text{ F}$

Thus, $X_C = \dfrac{0.159}{10^3 \times 10^{-6}} = 160 \text{ ohms}$

15. Now do these two simple examples. In each case find the capacitor's reactance at 1 kHz (X_{C1}) and at the given frequency (X_{C2}).

(a) $C = 0.1 \mu\text{F}$, $f = 100 \text{ Hz}$. _____

(b) $C = 100 \mu\text{F}$, $f = 2 \text{ kHz}$. _____

- - - - - - - - - - - - - - - - -

(a)   at 1 kHz, $X_{C1} = 1600 \text{ ohms}$; at 100 Hz, $X_{C2} = 16{,}000 \text{ ohms}$
(b)   at 1 kHz, $X_{C1} = 1.6 \text{ ohms}$; at 2 kHz, $X_{C2} = 0.8 \text{ ohms}$

A capacitor in series with a resistor will make a voltage divider, just as will two resistors.

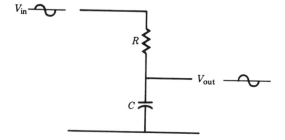

The pp voltage out will be less than the pp voltage in. However the relationship is not as simple as that. If the output and input voltage waveforms were placed on the same time axis, we would see that one is shifted from the other. The two waveforms are said to be "out of phase." *Phase* is an important concept in understanding how certain electronic circuits work. In the next chapter we will be discussing phase relationships for some AC circuits. We will also see it again when studying amplifiers.

THE INDUCTOR IN AN AC CIRCUIT

16. An inductor is a coil of wire, usually many turns around a piece of soft iron. In some cases the wire is wound around a nonconducting material.

(a)   Is the AC reactance of an inductor high or low? Why? _____

_____

(b)   Is the DC resistance high or low? _____

(c)   What is the relationship between the AC reactance and the DC

resistance? _____

(d)   What is the formula for the reactance of an inductor?

_____

– – – – – – – – – – – – – – – – –

(a)   Its AC reactance, which can be quite high, is caused by the magnetic effects of the changing current through the coils.
(b)   usually quite low
(c)   none
(d) $X_L = 2\pi fL$, where L = the value of the inductance in henrys.

17.   Assume the inductance value is 10 henrys (H) and the frequency is
100 Hz. Find the reactance.

_____

– – – – – – – – – – – – – – – – –

$X_L = 2\pi fL = 2\pi \times 100 \times 10 = 6280$ ohms

18.   Now try these two examples. In each case find the reactance of the
inductor at 1 kHz (X_{L1}) and at the given frequency (X_{L2}).

(a) $L = 1$ mH (0.001 H), $f = 10$ kHz. _____

(b) $L = 0.01$ mH, $f = 5$ MHz. _____

– – – – – – – – – – – – – – –

(a) $X_{L1} = 6.28 \times 10^3 \times 0.001 = 6.28$ ohms
$X_{L2} = 6.28 \times 10 \times 10^3 \times 0.001 = 62.8$ ohms
(b) $X_{L1} = 6.28 \times 10^3 \times 0.01 \times 10^{-3} = 0.0628$ ohms
$X_{L2} = 6.28 \times 5 \times 10^6 \times 0.01 \times 10^{-3} = 314$ ohms

As with the capacitor, the inductor will also make a voltage divider
when placed in series with a resistor. Again, the relationship between the
input and output voltages is not as simple as a resistive divider. The circuit
will be discussed in Chapter 6.

RESONANCE

19.   Previous frames have shown that the capacitive reactance decreases
as the frequency increases and that inductive reactance increases as
frequency increases. If a capacitor and an inductor are connected in
series there will be one frequency at which their reactances are equal.

(a) What is this frequency called? _____

(b) What is the formula for calculating the resonant frequency? You can find it by setting $X_L = X_C$ and solving for frequency.

_____

- - - - - - - - - - - - - - - -

(a) the resonant frequency
(b) $2\pi fL = 1/(2\pi fC)$ yields the following formula:

$$f = \frac{1}{2\pi\sqrt{LC}}$$

20. If a capacitor and an inductor are connected in parallel there will also be a resonant frequency. Analysis of a parallel resonant circuit is not as simple as for a series resonant circuit. The reason for this is that inductors always have some internal resistance, which complicates some of the equations. However, under certain conditions the analysis is similar. For example, if the reactance of the inductor in ohms is more than 10 times greater than its own internal resistance, the formula for the resonant frequency is the same as if the inductor and capacitor were connected in series. This is an approximation that we will often use.

For the following inductors, determine if the reactance is more or less than 10 times its internal resistance. A resonant frequency is given.

(a) $f = 25$ kHz, $L = 2$ mH, $R_{in} = 20$ ohms _____

(b) $f = 1$ kHz, $L = 33.5$ mH, $R_{in} = 30$ ohms _____

- - - - - - - - - - - - - - - -

(a) $X_L = 314$ ohms, which is more than 10 times greater than R_{in}.
(b) $X_L = 210$ ohms, which is less than 10 times greater than R_{in}.

Note: In Chapter 7 we will be discussing both the series and parallel resonant circuits. At that time you'll learn many techniques and formulas to aid in your study.

21. Find the resonant frequency (f_r) for the capacitors and inductors below when they are connected both in parallel and in series. Assume R_{in} is negligible.

(a) $C = 1\ \mu F$, $L = 1$ henry. _____

(b) $C = 0.2\ \mu F$, $L = 3.3$ mH. _____

- - - - - - - - - - - - - - - -

(a) $f_r = 0.159/\sqrt{10^{-6} \times 1} = 160$ Hz
(b) $f_r = 0.159/\sqrt{3.3 \times 10^{-3} \times 0.2 \times 10^{-6}} = 6.2$ kHz

22. Now try these two final examples.

(a) $C = 10\ \mu F$, $L = 1$ henry. _____

(b) $C = 0.0033\ \mu F$, $L = 0.5$ mH. _____

- - - - - - - - - - - - - - - -

(a) $f = 50$ Hz (approximately)
(b) $f = 124$ kHz

Resonance is very important in electronics. Two areas of practical application found most frequently are in filters and oscillators.

Filters are attenuators which will attenuate only a certain band of frequencies and not others. They are very important in radio, TV, and other communications. Oscillators are electronic circuits which generate a continuous output without needing an input signal. The type of oscillator which uses a resonant circuit produces pure sine waves. (We will learn more about oscillators in Chapter 9.)

SUMMARY

The concepts presented in this chapter can be summarized as follows:

- The sine wave is used extensively in AC circuits.

- The most common laboratory generator is the function generator.

- $V_p = \sqrt{2} \times V_{rms}$, $V_{pp} = 2\sqrt{2} \times V_{rms}$

- $f = 1/T$

- $I_{pp} = \dfrac{V_{pp}}{R}$, $I_{rms} = \dfrac{V_{rms}}{R}$

- $X_C = \dfrac{1}{(2\pi f C)}$, capacitive reactance

- $X_L = 2\pi f L$, inductive reactance

- $f = \dfrac{1}{2\pi\sqrt{LC}}$, resonant frequency

SELF-TEST

The following problems will test your understanding of the basic concepts in this chapter. Use a separate sheet of paper if necessary. Compare your answers with the answers provided following the test.

1.   Convert the following peak or peak-to-peak values to RMS values:

   (a) $V_p = 12$ V, $V_{rms} =$ _____

   (b) $V_p = 80$ mV, $V_{rms} =$ _____

   (c) $V_{pp} = 100$ V, $V_{rms} =$ _____

2.   Convert the following RMS values to the required values shown:

   (a) $V_{rms} = 120$ V, $V_p =$ _____

   (b) $V_{rms} = 100$ mV, $V_p =$ _____

   (c) $V_{rms} = 12$ V, $V_{pp} =$ _____

3.   For the given value, find the period or frequency:

   (a) $T = 16.7$ ms, $f =$ _____

   (b) $f = 15$ kHz, $T =$ _____

4.   For the circuit shown, find the total current flow and the voltage across R_2, (V_{out}).

$V_{in} = 20$ V_{rms}

80Ω

120Ω V_{out}

5.   Find the reactances of the following components:

   (a) $C = 0.16$ μF, $f = 12$ kHz, $X_C =$ _____

   (b) $L = 5$ mH, $f = 30$ kHz, $X_L =$ _____

6.   Find the frequency necessary to cause the reactances shown:

   (a) $C = 1\mu$F, $X_C = 200$ ohms, $f =$ _____

   (b) $L = 50$ mH, $X_L = 320$ ohms, $f =$ _____

7.   What would be the resonant frequency for the capacitor and inductor values given in (a) and (b) of problem 5 if they were connected in series? _____

8.   What would be the resonant frequency for the capacitor and inductor values given in (a) and (b) of problem 6 if they were connected in parallel? What assumption would you be making? _____

_____

_____

Answers to Self-Test

If your answers do not agree with those given below, review the frames
indicated in parentheses before you go on to the next chapter.

1.   (a) 8.5 V_{rms}
     (b) 56.6 mV_{rms}
     (c) 35.4 V_{rms} (frames 4–6)

2.   (a) 169.7 V_p
     (b) 141.4 mV_p
     (c) 33.9 V_{pp} (frames 4–6)

3.   (a) 60 Hz
     (b) 66.7 μsec (frame 7)

4. $I_T = 0.1\ A_{rms}$, $V_{out} = 12\ V_{rms}$ (frames 9–11)

5.   (a) 82.9 ohms
     (b) 942.5 ohms   (frames 14 and 17)

6.   (a) 795.8 Hz
     (b) 1.02 kHz   (frames 14 and 17)

7.   5.63 kHz   (frame 19)

8.   711.8 Hz. Assume the internal resistance of the inductor is negligible.
                                                             (frame 20)

CHAPTER SIX

AC in Electronics

After the basics of AC theory, certain other topics are essential to the study of electronic circuits. The most useful of these, voltage dividers formed by capacitors and resistors, and inductors and resistors, are covered in this chapter. These concepts and their effects on AC signals play a major part in communications, consumer items, and industrial controls. For example, a type of circuit shown here is used to eliminate the familiar 60 Hz hum in audio circuits, as well as to provide the simple "tone controls" on hi-fi equipment.

When you complete this chapter you will be able to:

- calculate the output voltage of a capacitor and resistor in series in an AC circuit;

- calculate the output voltage of a capacitor and resistor in parallel in an AC circuit;

- calculate the effects of a simple parallel RC voltage divider;

- determine the AC attenuation in a parallel RC voltage divider;

- draw the output waveform of an AC or combined AC-DC voltage divider circuit;

- calculate simple phase angles and phase differences;

- use inductive reactance in voltage divider output calculations.

CAPACITORS IN AC CIRCUITS

1. An AC signal is continually changing, whether it is a pure sine wave or a complex wave made up of many sine waves. If such a signal is applied to one plate of a capacitor, it will be observed at the other plate. To express this another way: a capacitor will "pass" an AC signal.

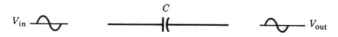

A capacitor does *not* look like an open circuit with an AC signal, as it does with a DC signal. Equally important: a capacitor is *not* a short circuit to an AC signal.

(a) What is the main difference observed in capacitor behavior when it is in an AC circuit instead of a DC circuit? _____

_____

(b) Does a capacitor appear as a short or an open circuit to an AC signal? _____

— — — — — — — — — — — — — — —

(a) It will pass an AC signal while it will not pass a DC voltage level.
(b) It is neither.

2. A capacitor will, in general, oppose the flow of an AC current. This opposition to current flow, as we saw in Chapter 5, is called the *reactance* of the capacitor.

  Reactance is similar to resistance, but is a little more complicated than this simple comparison suggests. However, we will not go deeper into it in this book. Reactance is expressed by a well known formula. Write the formula for the reactance of a capacitor.

_____

— — — — — — — — — — — — — — —

$$X_c = \frac{1}{2\pi fC}$$

3. From this formula it is easy to see that the reactance changes when the frequency of the input signal changes. If the frequency increases, what happens to the reactance? _____

— — — — — — — — — — — — — — —

It decreases.

  If you had difficulty with these first three frames, you should review the examples worked in Chapter 5.

CAPACITORS AND RESISTORS IN SERIES

4. For simplicity we will consider all inputs at this time to be pure sine waves. The circuit shown at the top of page 122 shows a sine wave input to a capacitor.

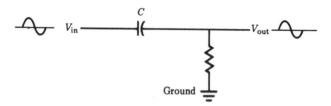

If the input is a pure sine wave, what is the output? _____

_ _ _ _ _ _ _ _ _ _ _ _ _ _ _ _

a pure sine wave

5. The output sine wave will have the same frequency as the input sine wave.
    A capacitor cannot change the frequency of the signal. But, as we said
    before, with an AC input the capacitor behaves in a manner similar to a
    resistor. It is *not* a short circuit.
        What can change in the above circuit is the amplitude of the output
    sine wave. The output amplitude will be different from the input in most
    cases.
        With an AC input to a simple circuit like the one above, what does

    the capacitor appear to behave like? _____

    _ _ _ _ _ _ _ _ _ _ _ _ _ _ _ _

    It appears to have opposition similar to a resistor.

6. The capacitor and resistor will also behave like a voltage divider.

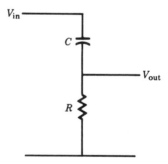

    In the case of two resistors forming a voltage divider, recall the formula

    for the voltage output. What is this formula? _____

    _ _ _ _ _ _ _ _ _ _ _ _ _ _ _ _

$$V_{out} = V_{in} \times \frac{R_2}{R_1 + R_2}$$

7. The two resistors in series will also present a total resistance to the flow of current. What is the formula for this total resistance?

_____

– – – – – – – – – – – – – –

$$R_T = R_1 + R_2$$

8. In the case of the capacitor and the resistor, this combination will also present an opposition to the flow of current. Here the opposition is called the *impedance* and it is expressed by the following formula.

$$Z = \sqrt{X_C^2 + R^2}$$

where: Z = the impedance of the circuit in ohms
X_C = the reactance of the capacitor in ohms
R = the resistance in ohms

To gain a little familiarity with this formula, calculate the impedance and the current in this circuit.

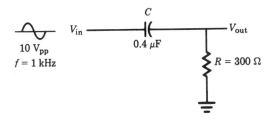

C

V_{in} ———— V_{out}

0.4 μF

10 V_{pp}
$f = 1$ kHz

$R = 300\ \Omega$

(a) $X_C = \dfrac{1}{2\pi f C} = $ _____

(b) $Z = \sqrt{X_C^2 + R^2} = $ _____

(c) $I = \dfrac{V}{Z} = $ _____

– – – – – – – – – – – – – –

(a) 400 ohms
(b) 500 ohms
(c) a sine wave with 20 mA$_{pp}$

Note: The answer for (c) is not the complete answer. Later in this chapter the phase shift nature of the circuit is discussed.

9. Now, using the same circuit, calculate the impedance and current for the next two examples.

   (a) $f = 60$ Hz, $C = 530 \mu F$, $R = 12$ ohms, $V_{in} = 26 V_{pp}$

   _____

   (b) $f = 10$ kHz, $C = 1.77 \mu F$, $R = 12$ ohms, $V_{in} = 150 V_{pp}$

   _____

   – – – – – – – – – – – – – – – –

   (a) $Z = 13$ ohms, $I = 2 A_{pp}$
   (b) $Z = 15$ ohms, $I = 10 A_{pp}$

10. As the resistor and-capacitor also form a voltage divider, it should be possible to calculate the voltage out.

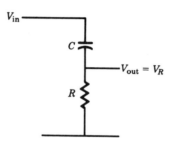

The formula is shown below.

$$V_{out} = V_{in} \times \frac{R}{Z}$$

Calculate the output voltage in this circuit.

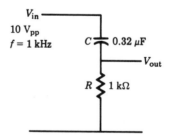

(a) Find X_C. _____

(b) Find Z. _____

(c) Use the formula to find V_{out}. _____

– – – – – – – – – – – – – – –

(a) $X_C = 500$ ohms (rounded off)
(b) $Z = 1120$ ohms (rounded off)
(c) $V_{out} = 8.9$ V$_{pp}$

11.   Now find V_{out} for these two examples.

(a) $V_{in} = 10$ V$_{pp}$, $f = 1$ kHz, $C = 0.16$ μF, $R = 1$ kΩ

(b) $V_{in} = 10$ V$_{pp}$, $f = 1$ kHz, $C = 0.08$ μF, $R = 1$ kΩ

— — — — — — — — — — — — — —

(a) $V_{out} = 7.1$ V$_{pp}$
(b) $V_{out} = 4.5$ V$_{pp}$

Note: Hereafter, it will be understood that the answer is a peak-to-peak value if the given value is a peak-to-peak value.

12.   The output voltage is said to be "attenuated" in the voltage dividers shown in frame 10. Compare the input and output voltages for each example.

What does "attenuated" mean? ——————————————

——————————————————————

— — — — — — — — — — — — — —

to reduce in amplitude or magnitude, that is, to make smaller

13.   In calculating V_{out} in the examples in frames 10–11, it was first neces-sary to find X_C. However, X_C changes as the frequency changes, while the resistance remains constant. So as the frequency changes, the imped-ance Z will change, and so also will the amplitude of the output voltage V_{out}.
        If V_{out} is plotted against frequency on a graph, the curve will look like that shown below.

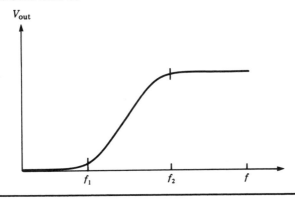

The frequency f_1 at which the curve starts to rise and f_2 where it starts to level off will both depend on the values of the capacitor and the resistor. The slope of the curve does not change; it is constant.

As an exercise, calculate the output voltage in this circuit for frequencies of 100 Hz, 1 kHz, 10 kHz, and 100 kHz.

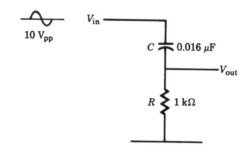

(a) 100 Hz. _____

(b) 1 kHz. _____

(c) 10 kHz. _____

(d) 100 kHz. _____

(e) Plot these values for V_{out} against f and draw an approximate curve.

- - - - - - - - - - - - - - - - - -

(a) $V_{out} = 0.1$ V
(b) $V_{out} = 1$ V
(c) $V_{out} = 7.1$ V
(d) $V_{out} = 10$ V
(e)

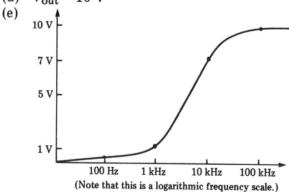

(Note that this is a logarithmic frequency scale.)

VOLTAGE DIVIDER EXPERIMENT

14. Set up the circuit shown in frame 13 and let the frequency vary as
     shown in the following table. Each time the frequency is changed, be
     sure that the input voltage amplitude is held constant at the same
     voltage. Sometimes the voltage output of a generator will change
     slightly as the frequency is varied. For each setting of the generator,
     calculate the capacitive reactance and the magnitude of the impedance.
     The experiment can be done using RMS values as measured by a
     voltmeter or pp values as measured by an oscilloscope.
          Again, plot the output voltage versus the frequency. Check to see
     how the measured values compared to the calculated values of frame
     13. The curve should have the same shape as the curve for frame 13 but
     may be noticed as being shifted slightly to the right or left.

f_{in}	X_C	Z	V_{out}
25 Hz			
50 Hz			
100 Hz			
250 Hz			
500 Hz			
1 kHz			
3 kHz			
5 kHz			
7 kHz			
10 kHz			
20 kHz			
30 kHz			
50 kHz			
100 kHz			

What would you expect to cause the curve to be moved slightly to

the right or the left? _____

_____

— — — — — — — — — — — — — — — —

the inexact values of practical components and the inaccuracies in the
frequency/voltage settings

The values obtained should be close to those shown in the table below, and the curve should be very similar to that at the bottom of this page.

f_{in}	X_C	Z	V_{out}
25 Hz	400 kΩ	400 kΩ	0.025 V
50 Hz	200 kΩ	200 kΩ	0.05 V
100 Hz	100 kΩ	100 kΩ	0.1 V
250 Hz	40 kΩ	40 kΩ	0.25 V
500 Hz	20 kΩ	20 kΩ	0.5 V
1 kHz	10 kΩ	10 kΩ	1 V
3 kHz	3.3 kΩ	3.5 kΩ	2.9 V
5 kHz	2 kΩ	2.2 kΩ	4.47 V
7 kHz	1.4 kΩ	1.7 kΩ	5.8 V
10 kHz	1 kΩ	1.414 kΩ	7.1 V
20 kHz	500 Ω	1.12 kΩ	8.9 V
30 kHz	330 Ω	1.05 kΩ	9.5 V
50 kHz	200 Ω	1.02 kΩ	9.8 V
100 kHz	100 Ω	1 kΩ	10 V

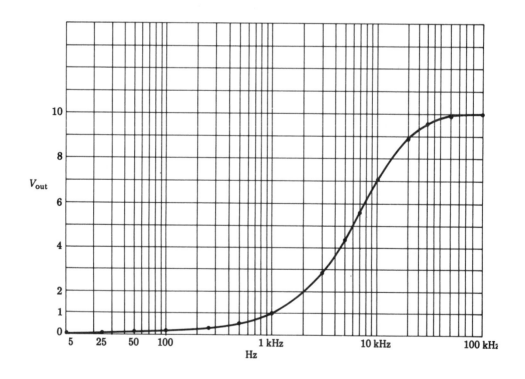

15. In AC work, the circuit below is used on many occasions.

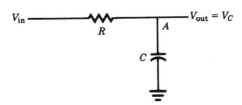

Here the output voltage is the voltage across the capacitor instead of the resistor (that is, measured between point A and ground).

The impedance of the circuit remains the same as before. It will still behave like a voltage divider, and the output voltage can be calculated as before.

(a) What is the impedance formula for the circuit? _____

(b) What is the formula for the output voltage? _____

- - - - - - - - - - - - - - -

(a) $Z = \sqrt{X_C^2 + R^2}$

(b) $V_{out} = V_{in} \times \dfrac{X_C}{Z}$

16. Use the circuit of frame 15 again, with the following values.

$$V_{in} = 10 \ V_{pp}, \ f = 2 \ kHz, \ C = 0.1 \ \mu F, \ R = 1 \ k\Omega$$

Find:

(a) X_C. _____

(b) Z. _____

(c) V_{out}. _____

- - - - - - - - - - - - - - -

(a) 795 ohms
(b) 1277 ohms approximately
(c) 6.24 V approximately

17. Now calculate the voltage across the resistor in the circuit in frame 16.

_____

- - - - - - - - - - - - - - -

$$V_R = V_{in} \times \frac{R}{Z} = 10 \times \frac{1000}{1277} = 7.83 \ V_{pp}$$

18. What should be the relationship between V_{in} and the voltages across the capacitor and the resistor? Check to see if it holds in this case.

_____

_ _ _ _ _ _ _ _ _ _ _ _ _ _ _ _ _ _

The relationship between the voltages is $V_{in}^2 = V_C^2 + V_R^2$.
In this example:
$$10^2 = 6.24^2 + 7.83^2$$
$$100 = 38.94 + 61.32 = 100.25$$
Thus the relationship is correct in this example.

Do not make the mistake of simply adding the voltages across the resistor and the capacitor. The squares of the voltages must be added to give the square of V_{in}.

19. The voltage output of the circuit is also dependent upon the frequency of the input, just as it was in the circuit introduced in frame 4. The graph of V_{out} against f looks like the following graph.

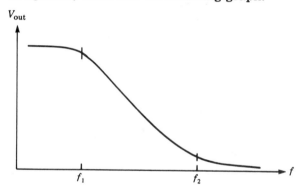

What values would you expect f_1 and f_2 to depend on? _____

_____

_ _ _ _ _ _ _ _ _ _ _ _ _ _ _ _ _ _

the capacitor and the resistor

PHASE SHIFT OF AN RC CIRCUIT

20. In both of the voltage divider circuits we have been considering, the output voltage is changed in amplitude from the input voltage.

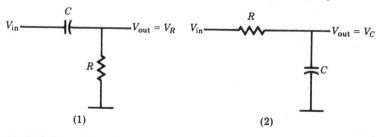

In what way is it changed? _____

- - - - - - - - - - - - - - - - - -

It is attenuated, or reduced.

21. The voltage is also changed in another way. In an AC circuit, the voltage
     across a capacitor rises and falls at the same frequency as the input, but
     it does not reach its peaks at the same time, nor does it pass through
     zero at the same time. This is shown in the following graphs.
          Graph 1 corresponds to figure 1 in frame 20. Graph 2 goes with
     figure 2.

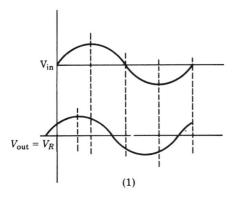

(1)

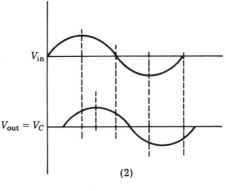

(2)

(a) Examine graph 1. Is the output voltage peak displaced to the right
     or the left? _____

(b) Examine graph 2. Is the output voltage peak displaced to the right
     or the left? _____

- - - - - - - - - - - - - - - - - -

(a) to the left
(b) to the right

22. The output voltage waveform in graph 1 is said to *lead* the input
     voltage waveform. In graph 2, the output waveform is said to *lag* the
     input waveform. This leading or lagging is measured in degrees found
     by observing the displacement between the peaks of the two
     waveforms, as shown in the graphs. The number of degrees can be
     approximated by remembering that a half cycle of a sine wave is 180
     degrees. It is important to note here that the difference between these
     two waveforms is called a *phase shift* or *phase difference*.

(a) What is the approximate phase shift of the two waveforms shown
     in the graphs? _____

(b) Do you think that the phase shift depends on the value of frequency? _____

(c) Will an RC voltage divider with the voltage taken across the capacitor produce a lead or a lag in the phase shift of the output voltage? _____

– – – – – – – – – – – – – – –

(a) approximately 35 degrees
(b) It does depend upon frequency since the value of the reactance and impedance depend upon frequency.
(c) a lag as shown in graph 2

23. The changes in current through a capacitor lead the voltage changes across the capacitor by 90 degrees. The current and voltage across a resistor are in phase, or have no phase difference.

Shown below is a vector diagram of a series RC circuit. θ is the phase angle by which V_R leads V_{in}. ϕ is the phase angle by which V_C lags V_{in}.

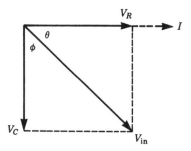

Note: Although the voltage across a resistor is in phase with the current through the resistor, both are out of phase with the applied voltage.

The phase angle is easily calculated. Because we must first determine $\tan \theta$ (the tangent of the phase angle) before θ (the phase angle), the calculation involves some knowledge of trigonometry. It is obtained from the formula below.

$$\tan \theta = \frac{V_C}{V_R} = \frac{1}{2 \pi f RC} = \frac{X_C}{R}$$

As an example, calculate the phase angle when 160 Hz is applied to a 3.9 kΩ resistor in series with a 0.1 μF capacitor.

$$\tan \theta = \frac{1}{160 \times 3.9 \times 10^3 \times 0.1 \times 10^{-6}} = 2.564$$

Using trigonometric tables, or a calculator, we would find that the phase angle is 68.7 degrees, which means that V_R leads V_{in} by 68.7 degrees. This also means that V_C lags the input by 21.3 degrees.

The types of diagram that show phase relationships as described opposite are actually called *phasor diagrams* in electronics, but the mathematics is the same as for vector diagrams. Students are generally more familiar with vector diagrams.

Sketch the phasor diagram for the calculations made in this frame.

– – – – – – – – – – – – – – – – –

It should look like this. Note that it also shows that the magnitude of V_C will be greater than V_R.

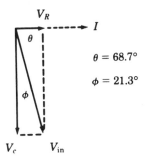

$\theta = 68.7°$

$\phi = 21.3°$

24. With the values shown in the circuit below, find the following.

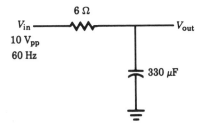

(a) X_C. _____

(b) Z. _____

(c) V_{out}. _____

(d) V_R. _____

(e) The current flowing. _____

(f) The phase angle. _____

– – – – – – – – – – – – – – –

(a) $X_C = \dfrac{1}{2\pi f C} = 8$ ohms

(b) $Z = \sqrt{8^2 + 6^2} = 10$ ohms

(c) $V_{out} = V_C = V_{in} \times \dfrac{X_C}{Z} = 8$ V

(d) $V_R = V_{in} \times \dfrac{R}{Z} = 6$ V

(e) $I = \dfrac{V}{Z} = 1$ amp

(f) $\tan \theta = \dfrac{X_C}{R} = \dfrac{8\,\Omega}{6\,\Omega} = 1.33$. Therefore, $\theta = 53.13°$

25. In the circuit below, calculate the same quantities as in frame 24.

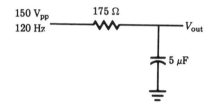

(a) X_C. _____

(b) Z. _____

(c) V_{out}. _____

(d) V_R. _____

(e) The current. _____

(f) The phase angle. _____

— — — — — — — — — — — — — — — —

(a) $X_C = 265$ ohms
(b) $Z = \sqrt{175^2 + 265^2} = 317.57\ \Omega$
(c) $V_C = 125$ V (approximately)
(d) $V_R = 83$ V (approximately)
(e) $I = 0.472$ A

(f) $\tan \theta = \dfrac{265\,\Omega}{175\,\Omega} = 1.5$ Therefore, $\theta = 56.56$ degrees.

RESISTOR AND CAPACITOR IN PARALLEL

26. The circuit below shows a common variation of the previous circuit.

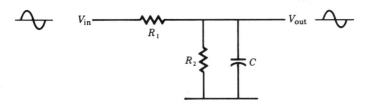

A DC input signal will not "see" the capacitor, and the circuit will function as shown below.

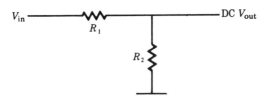

An AC input will "see" the capacitor, and it will see it in parallel with R_2. So the circuit will function as shown below, where r is the parallel equivalent of R_2 and X_C.

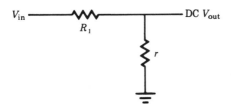

Calculating the exact parallel equivalent, r, is very complicated and will not be undertaken here. Instead, to demonstrate the usefulness of this circuit we will make a major simplification. We will look at the case where X_C is only about one tenth the value of R_2. This has many practicle applications as it will attenuate the AC and the DC differently.

An example with numbers will make the point clear. For the following circuit, you will calculate the AC and DC output voltages separately.

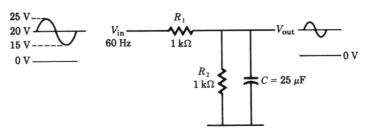

Follow the procedure below.

(a) Find X_C. Check that it is less than one tenth of R_2. $X_C = $ _____

(b) Look at the DC circuit. Use the voltage divider formula to find DC V_{out}.   DC $V_{out} = $ _____

(c) Look at the AC circuit. Use the voltage divider to find AC V_{out}.   AC $V_{out} = $ _____

(d) Compare the AC and DC input and output voltages. _____

_____

- - - - - - - - - - - - - - - -

(a) $X_C = 106$ ohms. $R_2 = 1000$ ohms, so X_C is close enough to one tenth of R_2.

(b) This is the DC circuit. $V_{out} = 20 \times \dfrac{1 \text{ k}\Omega}{1 \text{ k}\Omega + 1 \text{ k}\Omega} = 10 \text{ V}$

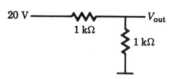

(c) This is the AC circuit. $V_{out} = 10 \times \dfrac{106}{\sqrt{(1000)^2 + (106)^2}} = 1.05 \text{ V}$

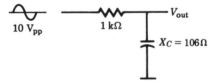

(d) The input and output waveforms are both drawn on the following figures. Note that the DC voltage has dropped from 20 V to 10 V, and that the AC voltage has dropped from 10 V to 1.05 V.

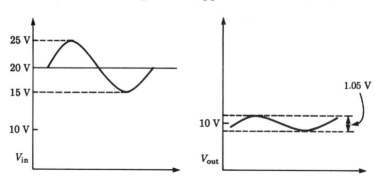

27. Here are two more examples. Use the same four steps to find and compare the output voltages with the input voltages. The input voltage is the same as in frame 26.

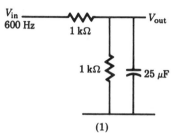

(1)

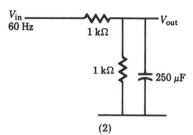

(2)

(1) (a) $X_C =$ _____

(b) DC $V_{out} =$ _____

(c) AC $V_{out} =$ _____

(d) Attenuation.

_____

(2) (a) $X_C =$ _____

(b) DC $V_{out} =$ _____

(c) AC $V_{out} =$ _____

(d) Attenuation.

_____

- - - - - - - - - - - - - - - -

(1) (a) $X_C = 10.6$ ohms
(b) DC $V_{out} = 10$ V
(c) AC $V_{out} = 0.1$ V (approximately)
(d) Here the DC attentuation is the same as the example in frame 26, but the AC has been reduced much more as a result of the frequency being increased.

(2) (a) $X_C = 10.6$ ohms
(b) DC $V_{out} = 10$ V
(c) AC $V_{out} = 0.1$ V
(d) The DC attentuation is still the same, but the AC has again been reduced. Note this time the larger capacitor was the cause.

These examples were made very simple by using X_C about one tenth of R_2. The answers produced are approximate, but are typical of what is found in practice when a simple RC filter is made to get rid of audio hum or some other unwanted frequency.

If X_C is larger or near R_2 in value the situation is very different and much more complex. In all cases the phase of the output changes and the amplitude difference is much more uncertain. To explore this fully requires a more advanced treatment and is beyond the scope of this book.

INDUCTORS IN AC CIRCUITS

28. An inductor can form a voltage divider with a resistor, as in the figure below.

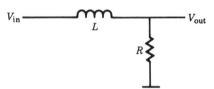

As before, we will consider all the inputs to be pure sine waves. Like the capacitor, the inductor cannot change the frequency of the sine wave, but it can affect the amplitude of the output voltage.

The simple series connection, as diagrammed above, will oppose current flow.

(a) What is the opposition to current flow called? _____

(b) What is the formula for the reactance of the inductor?

_____

(c) Write out the formula for the opposition to the current flow for

this circuit. _____

- - - - - - - - - - - - - - - -

(a) impedance.
(b) $X_L = 2\pi f L$
(c) $Z = \sqrt{X_L^2 + R^2}$

In many cases the DC resistance of the inductor is very low and can often be assumed to be 0 ohms. Even when it is not 0 ohms, the resistance is often ignored since the value of the reactance X_L is often very high. We will assume the DC resistance is 0 ohms for now, but later there will be cases when it cannot be ignored.

29. In the circuit or voltage divider below, the voltage output is given by the voltage divider formula with the value of the inductor included.

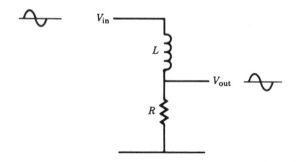

What is the formula for the voltage out? _____

- - - - - - - - - - - - - - - - -

$$V_{out} = V_{in} \times \frac{R}{Z}$$

30. Find the output voltage in the circuit shown below.

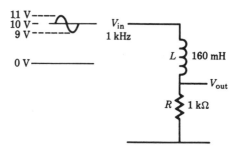

Follow the procedure.

(a) Find the DC output voltage. Use
     the DC voltage divider formula.   DC V_{out} = _____

(b) Find the reactance of the inductor. X_L = _____

(c) Find the AC impedance. Z = _____

(d) Find the AC output voltage.   AC V_{out} = _____

(e) Combine the outputs to find the actual output. Draw a simple
     graph.

- - - - - - - - - - - - - - - - - -

(a) DC $V_{out} = 10 \times \dfrac{1\ k\Omega}{1\ k\Omega + 0} = 10$ V

(b) $X_L = 1\ k\Omega$ (approximately)
(c) $Z = \sqrt{1^2 + 1^2} = \sqrt{2} = 1.414\ k\Omega$

(d) AC $V_{out} = 2 \times \dfrac{1\ k\Omega}{1.414\ k\Omega} = 1.414\ V_{pp}$

(e)

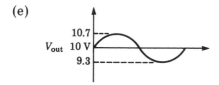

31. The circuit below illustrates a case where the DC resistance is not 0 ohms.

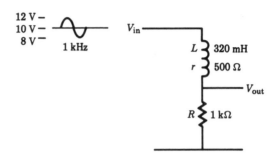

Follow the same procedure to find the output voltage. Repeat steps (a) through (e) from the previous frame.

(a) DC V_{out} = _____

(b) X_L = _____

(c) Z = _____

(d) AC V_{out} = _____

(e)

- - - - - - - - - - - - - - - -

(a) DC $V_{out} = \dfrac{10 \times 1k\Omega}{(1\ k\Omega + 500\Omega)} = 6.67$ V

Note that the 500 ohms internal resistance of the indicator will affect this voltage.

(b) $X_L = 2$ kΩ
(c) $Z = \sqrt{1.5^2 + 2^2} = 2.5$ kΩ
(d) AC $V_{out} = 1.6$ V$_{pp}$

(e)

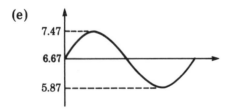

32. In calculating V_{out} in the examples in frames 30–31, it was necessary to know X_L. However, X_L changes with the frequency while the resistance remains constant.

Thus, as the frequency changes the impedance will also change, and so will the amplitude of V_{out}. If the output voltage V_{out} is plotted against the frequency it will look like that shown below.

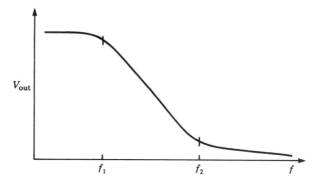

The frequency f_1 at which the curve starts to fall and f_2 where it begins to level off will both be dependent on certain values. The slope of the curve will be constant. What values will control f_1 and f_2?

_____

– – – – – – – – – – – – – – – –

the inductor and the resistor

33. The resistor and the inductor can also be set up this way, with the output voltage taken across the inductor.

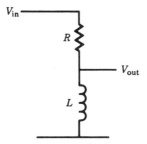

(a) What formula would be used to find V_{out}? _____

(b) What shape would you expect for the V_{out}–f curve?

– – – – – – – – – – – – – – – –

(a) $V_{out} = V_{in} \times \dfrac{X_L}{Z}$

(b)

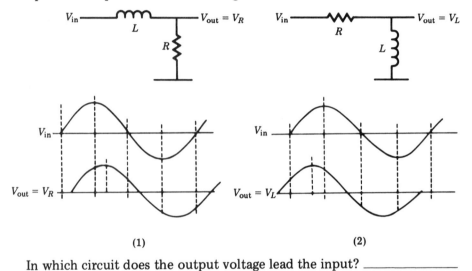

PHASE SHIFT FOR AN RL CIRCUIT

34. As with the capacitor, the inductor will also produce a phase shift in the output signal. As before, the phase shift or phase difference between the input and output is measured in degrees.

(1)   (2)

In which circuit does the output voltage lead the input? _____

In graph 1 the output voltage lags the input, and in graph 2 it can be seen to lead.

35. A vector diagram can be drawn for both the circuits in frame 34. The current through the inductor lags the voltage across the inductor by 90 degrees.

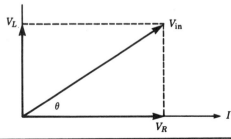

The phase angle is easily found.

$$\tan \theta = \frac{V_L}{V_R} = \frac{X_L}{R} = \frac{2\pi f L}{R}$$

Refer back to the problem you worked in frame 30. What is the phase

angle? _____

——————————————

45 degrees

36. Refer back to frame 31. What is the phase angle? _____

——————————————

$$\tan \theta = \frac{X_L}{R} = \frac{2\ k\Omega}{1.5\ k\Omega} = 1.33,\ \text{therefore}\ \theta = 53.1\ \text{degrees}$$

SELF-TEST

The questions below will test your understanding of this chapter. Use a
separate sheet of paper for your diagrams or calculations. Compare your
answers with the answers provided following the test.

In the circuits shown in questions 1–3, find the following quantities.

(a) X_C
(b) Z
(c) V_{out}

(d) I
(e) $\tan \theta$ and θ

1.

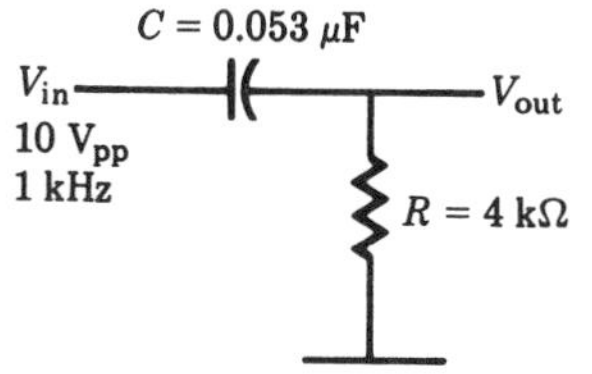

C = 0.053 μF
V_{in}
10 V_{pp}
1 kHz
V_{out}
$R = 4\ k\Omega$

(a) _____
(b) _____
(c) _____
(d) _____
(e) _____

2.

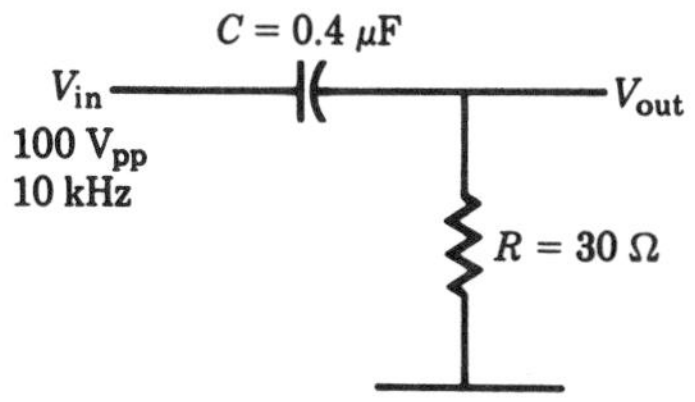

C = 0.4 μF
V_{in}
100 V_{pp}
10 kHz
V_{out}
$R = 30\ \Omega$

(a) _____
(b) _____
(c) _____
(d) _____
(e) _____

3.

$R = 12\ \Omega$

V_{in} ———ww——— V_{out}

26 V_{pp}
1 kHz

$C = 32\ \mu F$

(a) _____
(b) _____
(c) _____
(d) _____
(e) _____

In questions 4–6, find the following quantities.

(a) X_C (c) DC V_{out}
(b) AC V_{out}

4.   AC 10 V_{pp}

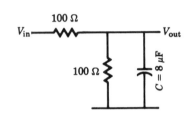

100 Ω

V_{in} ——ww—— V_{out}

20 V

2 kHz   100 Ω $C = 8\ \mu F$

(a) _____   (c) _____

(b) _____

5.   AC 15 V_{pp}

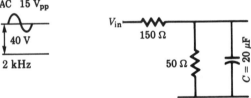

V_{in} ——ww——

40 V   150 Ω

2 kHz   50 Ω $C = 20\ \mu F$

(a) _____   (c) _____

(b) _____

6.   AC 10 V_{pp}

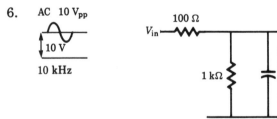

100 Ω

V_{in} ——ww——

10 V

10 kHz   1 kΩ $C = 0.25\ \mu F$

(a) _____   (c) _____

(b) _____

In questions 7–9, find the following quantities.

(a) DC V_{out} (d) AC V_{out}
(b) X_L (e) tan θ and θ
(c) Z

7. AC 3.13 V_{pp}

10 V

1 kHz

$L = 0.48$ mH

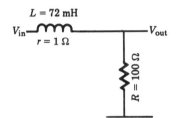

V_{in} V_{out}
$r = 1\,\Omega$

$R = 9\,\Omega$

(a) _____   (d) _____

(b) _____   (e) _____

(c) _____

8. AC 9.1 V_{pp}

10 V

2 kHz

$L = 72$ mH

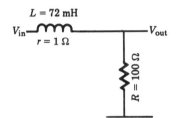

V_{in} V_{out}
$r = 1\,\Omega$

$R = 100\,\Omega$

(a) _____   (d) _____

(b) _____   (e) _____

(c) _____

9. AC 10 V_{pp}

10 V

4 kHz

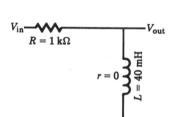

V_{in} V_{out}
$R = 1$ kΩ

$r = 0$ $L = 40$ mH

(a) _____   (d) _____

(b) _____   (e) _____

(c) _____

Answers to Self-Test

If your answers do not agree with those given below, review the applicable frames in the chapter before you go on to the next chapter.

1.   (a)   3 kΩ   (d)   2 A
      (b)   5 kΩ   (e)   36.87 degrees
      (c)   8 V

2.   (a)   40 ohms   (d)   2 A
      (b)   50 ohms   (e)   53.13 degrees
      (c)   60 V

3.   (a)   5 ohms   (d)   2 A
      (b)   13 ohms   (e)   22.63 degrees
      (c)   10 V

For problems 1, 2, and 3, see frames 8, 9, and 23.

4.   (a)   10 ohms   (c)   10 V
      (b)   1 V

5.   (a)   4 ohms   (c)   10 V
      (b)   0.4 V

6.   (a)   64 ohms   (c)   9.1 V
      (b)   5.4 V

For problems 4, 5, and 6, see frames 26 and 27.

7.   (a)   9 V   (d)   2.7 V
      (b)   3 ohms   (e)   16.7 degrees
      (c)   10.4 ohms

8.   (a)   10 V   (d)   1 V
      (b)   904 ohms   (e)   83.69 degrees
      (c)   910 ohms

9.   (a)   0 V   (d)   5 V
      (b)   1 kΩ   (e)   45 degrees
      (c)   1.414 kΩ

For problems 7, 8, and 9, see frames 28–30 and 34.

CHAPTER SEVEN

Resonant Circuits

We have seen how the inductor and the capacitor each present an opposition to the flow of an AC current, and how the magnitude of this reactance depends upon the frequency of the applied signal.

When these two components are used together, in series or in parallel, they have a definite and profound effect, and the peculiar phenomenon of *resonance* occurs.

In this chapter we will look at some of the properties of resonant circuits, and concentrate on those properties which lead to the study of oscillators. After completing this chapter you will be able to:

- find the impedance of a series *LC* circuit;

- calculate its resonant frequency;

- sketch a graph of its output voltage;

- find the impedance of a parallel *LC* circuit;

- calculate its resonant frequency;

- sketch a graph of its output voltage;

- calculate the bandwidth and the *Q* of a simple series and parallel circuit;

- find the frequency of an oscillator.

In this chapter, we will deal with AC only.

THE CAPACITOR AND INDUCTOR IN SERIES

1. In the simple circuit shown below we will look at the effects of a capacitor and an inductor placed in series.

This series combination can be used to form a voltage divider with a resistor. This is often called an RLC circuit since it contains R (resistance), L (inductance), and C (capacitance). The two circuits below act the same.

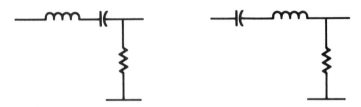

At this time we will assume the small DC resistance of the inductor is much less than the resistance of the resistor R, and can be safely ignored.

If an AC signal is applied to this circuit, both the inductor and the capacitor will have a reactance value which depends on the frequency.

(a)   What is the formula for the inductor's reactance? _____

(b)   What is the formula for the capacitor's reactance? _____

_ _ _ _ _ _ _ _ _ _ _ _ _ _ _ _ _

(a) $X_L = 2\pi f L$

(b) $X_C = \dfrac{1}{2\pi f C}$

2.   The series combination of the inductor and the capacitor will have a net reactance given by the formula below.

$$X = X_L - X_C$$

The impedance of the circuit is given by the following.

$$Z = \sqrt{R^2 + X^2}$$

In the circuit of frame 1, use the values below.

$f = 1$ kHz, $L = 100$ mH, $C = 1$ μF, $R = 500$ ohms

Find the net reactance and the impedance, using this procedure.

(a)   Find X_L. _____

(b)   Find X_C. _____

(c)   Use $X = X_L - X_C$ to find the net reactance. _____

(d)   Use $Z = \sqrt{X^2 + R^2}$ to find the impedance. _____

_ _ _ _ _ _ _ _ _ _ _ _ _ _ _ _

(a) $X_L = 628$ ohms
(b) $X_C = 160$ ohms

(c) X = 468 ohms (inductive)
(d) Z = 685 ohms

By convention X_C has a minus sign.

3. Now use the values below and the same procedure to find the net reactance and impedance.

f = 100 Hz, L = 0.5 H, C = 5 μF, R = 8 ohms

(a) X_L = _____

(b) X_C = _____

(c) X = _____

(d) Z = _____

- - - - - - - - - - - - - - - - - -

(a) X_L = 314 ohms
(b) X_C = 318 ohms
(c) X = −4 ohms (capacitive)
(d) Z = 9 ohms

4. Now find X and Z for these two examples.

(a) f = 10 kHz, L = 15 mH, C = 0.01 μF, R = 494 ohms

X = _____ Z = _____

(b) f = 2 MHz, L = 8 μH, C = 0.001 μF, R = 15 ohms

X = _____ Z = _____

- - - - - - - - - - - - - - - - - -

(a) X = 650 ohms capacitive, Z = 816 ohms
(b) X = 21 ohms inductive, Z = 25.8 ohms

5. If an AC signal is applied to this voltage divider, an output can be taken across the resistor.

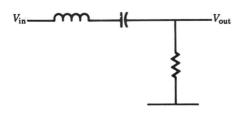

To calculate V_{out} for several frequencies is a long process. It is easier in practice to use several frequencies as input and to measure the output at each frequency. If the input amplitude is kept constant, then as the frequency increases the output voltage will rise and fall in amplitude as graphed below.

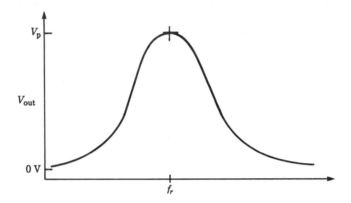

V_p, the maximum output voltage or peak voltage, will be approximately the same amplitude as V_{in}. The small resistance of the inductor will cause it to be slightly less than V_{in}.

What is important, and can be easily calculated, is the frequency at which the output peaks, f_r, which occurs when the net reactance is at a minimum.

Under ideal conditions, if X_C were 10.6 ohms, what value of X_L would result in the minimum impedance of that circuit, X_{min}, being 0?

— — — — — — — — — — — — — — —

10.6 ohms $(X_{min} = X_L - X_C = 0)$

6.   The frequency at which $X_L = X_C$ can be found using the two reactance formulas. When these are equal we have:

$$2\pi fL = \frac{1}{2\pi fC}$$

from which $f_r = \dfrac{1}{2\pi\sqrt{LC}}$

This is the *resonant frequency* of the circuit. What effect does the value of the resistance have on the resonant frequency? _____

— — — — — — — — — — — — — — —

It has no effect at all.

7. In these examples, the component values are given. Find the resonant frequency for each.

(a) $C = 1 \ \mu F$, $L = 1 \ mH$

$f_r =$ _____

(b) $C = 16 \ \mu F$, $L = 1.6 \ mH$

$f_r =$ _____

- - - - - - - - - - - - - - - - -

(a) $f_r = \dfrac{1}{2\pi\sqrt{1 \times 10^{-3} \times 1 \times 10^{-6}}} = 5.0 \ kHz$ (approximately)

(b) $f_r = \dfrac{1}{2\pi\sqrt{16 \times 10^{-6} \times 1.6 \times 10^{-3}}} = 1 \ kHz$ (approximately)

8. Calculate the resonant frequency for each of these circuits.

(a) $C = 0.1 \ \mu F$, $L = 1 \ mH$

$f_r =$ _____

(b) $C = 1 \ \mu F$, $L = 2 \ mH$

$f_r =$ _____

- - - - - - - - - - - - - - - - -

(a) $f_r = 16 \ kHz$ (approximately)
(b) $f_r = 3.6 \ kHz$ (approximately)

9. The series combination can be set up as shown to the right.

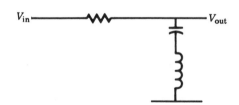

To see how the output voltage varies with frequency you would want to make measurements. The graph of the results looks like this.

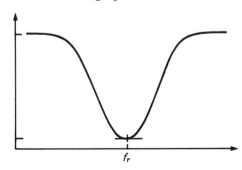

The minimum output again occurs when X is minimum, and this again is the resonant frequency. What would you expect this minimum output to be? _____

– – – – – – – – – – – – – – – –

0 V, or close to it

10. The capacitor and inductor can be connected in parallel, as shown here.

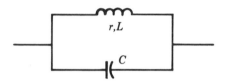

In this circuit, the internal resistance of the inductor has an effect on the resonance of the circuit and its operation with regard to frequency response. The formulas for this become much more complicated than for series resonance, and it is beyond the scope of this book to try to derive them. However, you can learn how to apply them to the study of resonance in a parallel circuit.

You will recall that in a series circuit the net reactance is equal to zero. In a parallel circuit, the reactances are inverted mathematically and then the sum of the inverses are equal to zero. In a series circuit at resonance, the net impedance is equal to the total resistance in the circuit including the resistance of the coil. In a parallel circuit it gets more complicated, as we shall see.

The resonance is found from the formula shown here.

$$f_r = \frac{1}{2\pi\sqrt{LC}}\sqrt{1 - \frac{r^2 C}{L}}$$

However, if the reactance of the coil is equal to or more than 10 times the resistance of the coil, then the formula for the series circuit can be used:

$$f_r = \frac{1}{2\pi\sqrt{LC}}$$

When we use this formula, we are saying that the Q of the coil is equal to or greater than 10 where $Q = X_L/r$. We will be using Q as a symbol from now on. It is a very useful approximation.

(a) Which formula will calculate the resonant frequency of a parallel circuit if the Q of the coil is 20?

_____

(b) If the Q is 8? _____

– – – – – – – – – – – – – – – –

(a)   the same formula as for the series circuit
(b)   the more complicated formula must be used

Note: Another version of this formula that is helpful when Q is known is shown here.

$$f_r = \frac{1}{2\pi\sqrt{LC}}\sqrt{\frac{Q^2}{1+Q^2}}$$

11.   The total opposition (impedance) of the parallel circuit to the flow of current can also be given by certain formulas at resonance.

$Z_p = Q^2 r$, if Q is equal to or greater than 10

$Z_p = \dfrac{L}{rC}$, for any value of Q

At resonance, the total impedance is considered to be all resistance, as the capacitive and inductive reactance effects have cancelled themselves out, just as they do in the resonant series circuit.
       The parallel combination can be used in a voltage divider with a resistor, as shown here.

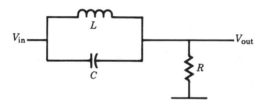

The voltage output curve is shown here and is most easily found by making measurements rather than calculations.

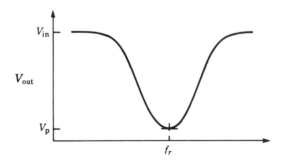

(a)   What would be the total impedance formula for the voltage divider circuit at resonance? _____

(b)   What is the frequency called at the point where the curve is at its minimum point? _____

(c) How would you interpret the curve at this minimum point? Why is the output voltage at a minimum value at resonance?

_____

_____

— — — — — — — — — — — — — — —

(a) $Z_T = Z_p + R$

Note: This is only true at resonance. At all other frequencies, Z_T is a complicated formula or calculation found by considering a series r,L circuit in parallel with a capacitor. A more detailed study of AC circuits would be necessary before we could develop this situation.

(b) the parallel resonant frequency
(c) The output voltage is at its lowest value at the resonant frequency. This is because the impedance of the parallel resonant circuit is at its highest value at this frequency.

12. In this circuit the output is taken across the parallel combination instead of the resistor.

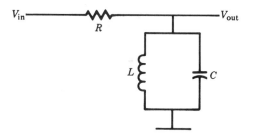

The output curve is shown here.

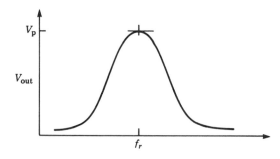

What formula would you expect to use to find the resonant frequency?

_____

— — — — — — — — — — — — — — — —

the same as before

13. Find the resonant frequency in these two examples, where the capacitor and the inductor are in parallel. (Q is greater than 10.)

(a) $L = 5$ mH, $C = 5$ μF

$f_r =$ _____

(b) $L = 1$ mH, $C = 10$ μF

$f_r =$ _____

- - - - - - - - - - - - - - - -

(a) $f_r = 1$ kHz (approximately)
(b) $f_r = 1600$ Hz (approximately)

THE OUTPUT CURVE

14. Now let us look at the output curve in a little more detail. The shape of this curve is very important. As an example we will use the curve below.

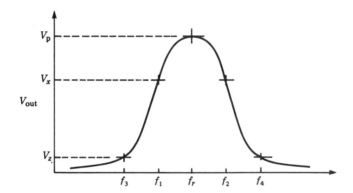

The resonant frequency, f_r, is a frequency which is "passed" by a circuit with this output curve. In other words, f_r appears at the output with minimum attenuation.

The two frequencies f_1 and f_2 are "passed" almost as well as f_r is passed. That is, they have a high output voltage, almost as high as the output of f_r. This voltage is shown on the graph as V_x.

The two frequencies f_3 and f_4 have a very low output voltage. These two are not passed, but are said to be "blocked" or "rejected" by the circuit. This voltage is shown on the graph as V_z.

The output or frequency response curve for a resonant circuit (series or parallel) has a symmetrical shape when Q is large (resistance of the coil is small). This is very important to note in the design of resonant circuits. The assumption is made that when Q is greater than 10, the output curve is symmetrical.

(a)   What is meant by a frequency which is passed? _____

_____

(b)   Why are f_1 and f_2 passed almost as well as f_r? _____

_____

(c)   What is meant by a frequency which is blocked? _____

_____

(d)   Which frequencies are blocked in the above circuit? _____

_____

(e)   Does the output curve shown appear to be symmetrical? What

does this mean with regard to the circuit? _____

_____

— — — — — — — — — — — — — — — —

(a)   It appears at the output with minimum attenuation.
(b)   because their output voltages are almost as high as that of f_r
(c)   It has a low output voltage.
(d) f_3 and f_4 (So also are all frequencies below f_3 and above f_4.)
(e)   It does appear to be symmetrical. This means that the coil has a
        Q greater than 10.

15.   Somewhere between f_r and f_3 and between f_r and f_4 there is a point at
      which frequencies are said to be passed or effectively blocked. This so-
      called dividing line is set at the level at which the power output of the
      circuit is half as much as it is at the peak value. This happens to occur
      at a level that is 0.707 or 70.7% of the peak value. For the output curve
      shown in frame 14, it would be at a voltage level of 0.707 V_p. The two
      corresponding frequencies taken from the graph are called the "half
      power frequencies" or "half power points." These are common expres-
      sions used in the design of resonant circuits and frequency response
      graphs.
          If a certain frequency results in an output voltage that is equal to
      or greater than the half power point, it is said to be passed or accepted
      by the circuit. If it is lower than the half power point, it is said to be
      blocked or rejected by the circuit.
          Knowing that Q is greater than 10 and what the peak value is,
      you can easily sketch the shape of the curve and identify the important
      frequencies. It is important to note that this method does allow us a
      way of comparing one circuit with another. In electronics, the shape of
      the output curve is very important and sometimes a particular shape is
      a design goal.

Suppose $V_p = 10$ V. What is the minimum voltage level of all frequencies which are passed by the circuit? _____

$V = 10$ V $\times$ 0.707 = 7.07 V (If a frequency has an output voltage above 7.07 V, it is said to be passed by the circuit.)

16. Assume the resonant frequency in a circuit is 5 V. Another frequency has an output of 3.3 V. Is this second frequency being passed or rejected through the circuit? _____

$V = V_p \times 0.707 = 5 \times 0.707 = 3.535$ V. 3.3 V is less than 3.535 V, so this frequency is being rejected.

17. In these examples, find the voltage level at the half power points.

(a) $V_p = 20$ V _____

(b) $V_p = 100$ V _____

(c) $V_p = 3.2$ V _____

(a) 14.14 V
(b) 70.70 V
(c) 2.262 V

18. Although we started off by talking about the center frequency of the curve, we have introduced a few other frequencies. So in fact we have become interested in a *band* or a *range* of frequencies.

There are two frequencies which correspond to the half power points on the curve. Let us now assume these are f_1 and f_2. The difference found by subtracting f_1 from f_2 is very important, as this gives the range of frequencies which are passed by the circuit. This range is called the *bandwidth* of the circuit, and is expressed by the equation below.

$$BW = f_2 - f_1$$

All frequencies outside this range are rejected.

Note that in some texts bandwidth is abbreviated by Δf.
Here we will use BW because it means only one thing—"bandwidth," rather than a variable value such as "change in frequency."

Indicate which of the following is the wider range of frequencies, that is, the wider bandwidth.

____ (a) $f_2 = 200$ Hz, $f_1 = 100$ Hz

____ (b) $f_2 = 20$ Hz, $f_1 = 10$ Hz

‒ ‒ ‒ ‒ ‒ ‒ ‒ ‒ ‒ ‒ ‒ ‒ ‒ ‒ ‒

bandwidth a is wider

When playing a radio, we are interested in hearing only one station at a time, not the adjacent stations. Thus the tuner must have a narrow bandwidth so that it can select the frequency of that one station only.

The amplifiers in a TV set, however, must pass frequencies from 30 Hz up to about 4.5 MHz, so a wider bandwidth is needed. Bandwidth required depends entirely upon the application or use to which the circuit will be put.

19.   In the last few frames we have used this curve.

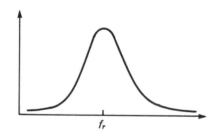

But everything we have said remains true for a curve with this shape.

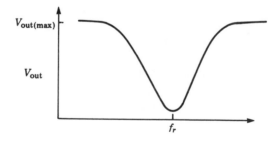

(a)   What points would you expect to determine the bandwidth for the second curve? _____

(b)   Would the resonant frequency be identified with a voltage that is above or below these points? _____

‒ ‒ ‒ ‒ ‒ ‒ ‒ ‒ ‒ ‒ ‒ ‒ ‒ ‒ ‒

(a)   the half power points (0.707 V$_{out}$)
(b)   The voltage level would be below the level for the half power
      points, which is the minimum point on the curve. In this case the
      points about the resonant frequency will be rejected.

20.   How "good" is a capacitor-inductor circuit at passing or rejecting a given
      frequency or a band of frequencies? In other words, although f_r is the
      main or center frequency, what other frequencies—wanted or unwanted
      —will be passed or rejected with f_r? This is the meaning of the bandwidth
      of the circuit.
           The bandwidth can be measured in practice or it can be calculated
      from the component values. It is very easy to calculate. Two formulas
      can be used.

$$\text{BW} = f_2 - f_1 \quad \text{or} \quad \text{BW} = \frac{f_r}{Q}$$

      where

$$Q = \frac{X_L}{R}$$

The Q of a circuit is a measure of its bandwidth.
      In a series circuit, R is the total DC resistance, including the
small DC resistance r of the inductor.

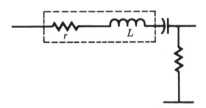

      With a parallel circuit Q refers only to the combination of the
inductor and the capacitor.

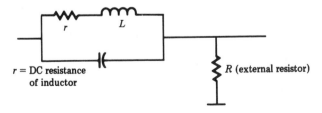

r = DC resistance
    of inductor

R (external resistor)

The external resistor R is not included in the computation of Q. *Only
the DC resistance r of the inductor is used.*

If an extra resistor is placed in either of the two places shown below, then it *is* included in the computation to find Q.

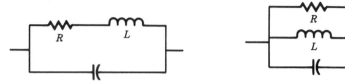

In the example circuit below, all the component values are given in the diagram. Find f_r, Q, and BW.

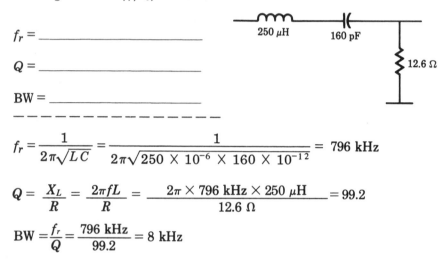

250 µH   160 pF   12.6 Ω

$f_r = $ _____

$Q = $ _____

BW = _____

- - - - - - - - - - - - - - - -

$$f_r = \frac{1}{2\pi\sqrt{LC}} = \frac{1}{2\pi\sqrt{250 \times 10^{-6} \times 160 \times 10^{-12}}} = 796 \text{ kHz}$$

$$Q = \frac{X_L}{R} = \frac{2\pi fL}{R} = \frac{2\pi \times 796 \text{ kHz} \times 250 \text{ µH}}{12.6 \text{ Ω}} = 99.2$$

$$BW = \frac{f_r}{Q} = \frac{796 \text{ kHz}}{99.2} = 8 \text{ kHz}$$

21. In the circuit diagrammed below find f_r, Q, and BW. Then draw an output curve showing the frequencies which are passed and rejected.

$f_r = $ _____

$Q = $ _____

BW = _____

10 mH   1 µF   100 Ω

- - - - - - - - - - - - - - - -

$f_r = 1590 \text{ Hz}$; $Q = 1$; BW $= 1590 \text{ Hz}$

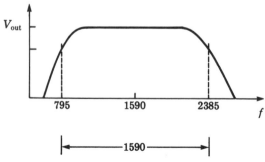

V_{out}

795   1590   2385 f

——1590——

22. Find f_r, Q, and the bandwidth for this circuit. Then draw the output curve.

$f_r =$ _____

$Q =$ _____

BW = _____

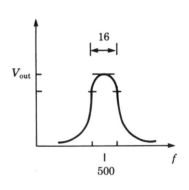

- - - - - - - - - - - - - - - -

$f_r = 500$ Hz; $Q = 31.4$, BW = 16 Hz

23. The circuit in this frame must be treated a bit differently. We require a resonant frequency of 1200 Hz and a Q of 80. R is given as 10 ohms. Find the bandwidth and the values of L and C required.

BW = _____

$L =$ _____

$C =$ _____

- - - - - - - - - - - - - - - -

BW = 15 Hz; $L = 106$ mH; $C = 0.166$ μF.

These values can be checked by using the values of L and C to find f_r.

24. We require a resonant frequency of 300 kHz with a bandwidth of 80 kHz. If $R = 10$ ohms, find the Q and the L and C values required.

$Q =$ _____

$L =$ _____

$C =$ _____

- - - - - - - - - - - - - - - -

$Q = 3.75$
$L = 20$ μH
$C = 0.014$ μF

25. A "high Q" circuit has a narrow bandwidth. It will only pass a very narrow band of frequencies. Its output curve looks like the one below.

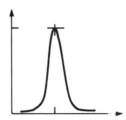

Because of the narrow range of frequencies it passes, the circuit is said to be very "selective" in the frequencies it passes.
   A "low Q" circuit, on the other hand, has a wide bandwidth. It is not very selective.

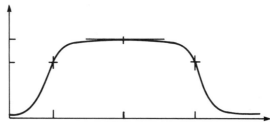

Recall the earlier discussion comparing the bandwidths of radio tuners and TV amplifiers.

(a) Which is the more selective, the radio tuner or the TV amplifier?

_____

(b) Which would require a lower Q circuit, the radio tuner or the TV amplifier? _____

- - - - - - - - - - - - - - - - - -

(a) the radio
(b) the TV amplifier

26. In this example the components are in parallel, but they are treated just as in the series circuit in frame 21. Find f_r, Q, and the bandwidth.

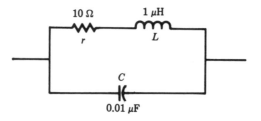

(a) $f_r =$ _____

(b) $Q =$ _____

(c)   BW = _____

- - - - - - - - - - - - - - -

(a) $f_r = 1.6$ MHz
(b) $X_L = 10$ ohms, so $Q = 10/0.01 = 100$ (Note that the only resis-
        tance here is the small DC resistance of the inductor.)
(c)   BW = 16 kHz (This is a fairly high Q circuit.)

27.   All of the examples in the last few frames have shown how to calculate
the BW and Q, or how to get a desired f_r and Q by finding needed component values.
        In each case a rough output curve could be sketched. However, it
is possible to draw a fairly accurate curve without recourse to extensive
calculations. The values given here will allow you to plot a curve accurate to within 1% of its true value. (These values can be proven mathematically, but it is a long process and will not be undertaken here.)
        The curve which results from this procedure is sometimes called
the *general resonance curve* and it is close enough for all practical work.
        To draw the curve make the following assumptions about certain
points on the curve.

(1)   The peak output voltage V_p at the resonant frequency f_r is
        assumed to be 100%. This is point A in the curve shown
        below.
(2)   The output voltage at f_1 and f_2 is 0.707 of 100% This is
        graphed as the two points labeled B. Note that $f_2 - f_1 = $ BW.
        In other words, at half a bandwidth above and below f_r the
        output is 70.7% of V_p.
(3)   At f_3 and f_4—points C—the output is 44.7% of V_p. Note that
        $f_4 - f_3 = 2$ BW. (In other words, at one bandwidth above and
        below f_r the output is 44.7% of maximum.)
(4)   At f_5 and f_6—points D—the output is 32% of V_p.
        Note $f_6 - f_5 = 3$ BW.
(5)   At f_7 and f_8—points E—the output is 24% of V_p.
        Note $f_8 - f_7 = 4$ BW.
(6)   At f_{10} and f_9—points F—the output is 13% of V_p.
        Note $f_{10} - f_9 = 8$ BW.

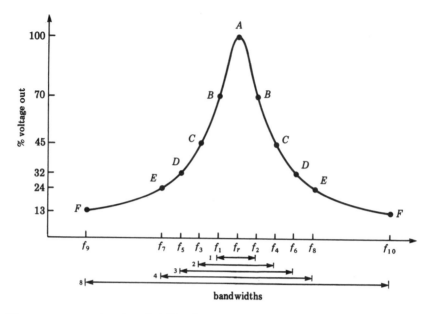

You are now going to plot the frequency curve for the circuit given here.

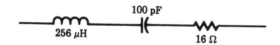

100 pF

256 μH

16 Ω

First calculate f_r, X_L, Q, and the BW.

$f_r =$ _____ $Q =$ _____

$X_L =$ _____   BW = _____

- - - - - - - - - - - - - - -

$f_r = 1$ MHz $Q = 100$

$X_L = 1607$ ohms   BW $= 10$ kHz

28. Now calculate the needed points for your graph. (Refer to the graph in frame 27 if necessary.)

(a) At what frequency will the output level be maximum? _____

(b) At what frequencies will the output level be 70% of V_p?

_____

(c) At what frequencies will the output level be 45% of V_p?

_____

(d) At what frequencies will the output level be 32% of V_p?

_____

(e) At what frequencies will the output level be 24% of V_p?

_____

(f) At what frequencies will the output level be 13% of V_p?

_____

— — — — — — — — — — — — — —

(a) 1 MHz
(b) 995 kHz and 1005 kHz (1 MHz − 5 kHz and + 5 kHz)
(c) 990 kHz and 1010 kHz
(d) 985 kHz and 1015 kHz
(e) 980 kHz and 1020 kHz
(f) 960 kHz and 1040 kHz

These points are all plotted on the following graph.

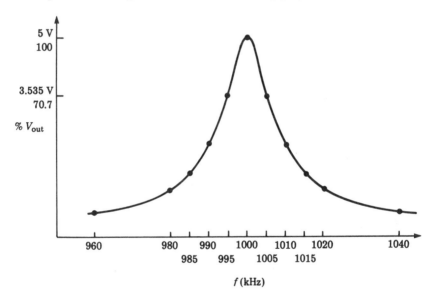

29. Answering the questions in the last frame allows you to plot the correct shape of the curve. All that now has to be done is to insert the correct voltage levels on the vertical axis.

You find the output voltage for the resonant frequency and put it at the 100% mark (V_p). Since you have percentages, you can easily find the other voltage levels.

For example, if the output voltage at f_r is 5 V, then the voltage output at the half power frequencies, f_1 and f_2, will be as calculated below.

$$V = 5 \text{ V} \times 70\% = 3.5 \text{ V (approximately)}$$

What is the voltage level at the 980 kHz frequency in the graph?

_____

_ _ _ _ _ _ _ _ _ _ _ _ _ _ _ _

$V = 5 \text{ V} \times 24\% = 1.2 \text{ V}$

INTRODUCTION TO OSCILLATORS

Inductive and capacitive circuits are widely used in situations where frequency selection or rejection is required. A few examples of this were mentioned earlier.

Another use is in oscillators, which we will study in more detail in Chapter 9. Many oscillators use a tuned parallel LC circuit to produce a sine wave output. The various circuit configurations used will be covered in Chapter 9, but here it is convenient to examine the action of the parallel resonant circuit and to explain briefly how it is used to make an oscillator.

30. In figure 1 below, when the switch is open no current flows and the capacitor is uncharged.

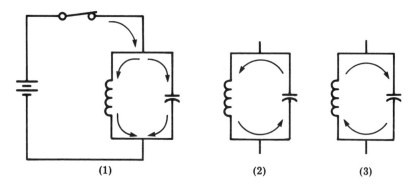

| (1) | (2) | (3) |

When the switch is closed, current flows through the completed circuit.
It is difficult for the current to flow through the inductor initially, as the inductor tends to oppose any changes in current flow. But it is easy for the current to flow into the capacitor and cause it to charge.
As the capacitor charges it passes less current, but now current is beginning to flow through the inductor. Eventually the capacitor is fully charged and a steady current flows through the inductor.
Now assume the switch is opened. What will happen to the

capacitor? _____

_ _ _ _ _ _ _ _ _ _ _ _ _ _ _ _

It discharges through the inductor. (Note the current direction, shown in figure 2.)

31. Since the current continues in the same direction through the inductor, it maintains the magnetic field. But the current decreases as the capacitor discharges. When the capacitor is fully discharged, how much current

    is flowing through the inductor? _____

    ― ― ― ― ― ― ― ― ― ― ― ― ― ―

    none

32. Since there is no current in the inductor the magnetic field collapses. As it collapses it induces a current to flow in the inductor, and this current flows in the *same* direction again, as shown in figure 2. This current now charges the capacitor in the reverse direction.

    When the magnetic field has fully collapsed, how much current

    will be flowing? _____

    ― ― ― ― ― ― ― ― ― ― ― ― ― ―

    none

33. Now the capacitor discharges through the inductor again, but this time the current flows in the opposite direction, as in figure 3. This builds a magnetic field of the opposite polarity. The magnetic field stops growing when the capacitor is fully discharged.

    Since there is now no current flowing the magnetic field collapses and induces current to flow in the direction shown in figure 3.

    What do you think this current will do to the capacitor? _____

    _____

    ― ― ― ― ― ― ― ― ― ― ― ― ― ―

    It charges it in the original direction.

34. When the field has fully collapsed the capacitor stops charging. It now begins to discharge again, causing current to flow through the inductor in the original direction shown in figure 2. This "see-saw" of current will continue indefinitely.

    As the current is flowing through the inductor the inductor will have a voltage drop across its ends. This voltage will vary as the current varies.

    What would you expect the voltage to look like when viewed on an

    oscilloscope? _____

    ― ― ― ― ― ― ― ― ― ― ― ― ― ―

    a sine wave

35. In a perfect circuit this will continue and a continuous sine wave will be
     produced. But in practice a small amount of power is lost in the DC
     resistance of the inductor and the other wiring. So the sine wave will
     gradually decrease in amplitude and die out to nothing after a few
     cycles.

How might this fade out be prevented? _____

_____

– – – – – – – – – – – – – – – – – –

by replacing a small amount of energy each cycle

This lost energy can be injected into the circuit by momentarily closing
and opening the switch at the correct time. (See figure 1 in frame 30.) This
would sustain the oscillations indefinitely.
       An electronic switch could be operated by the voltage variations across
the inductor, or by the voltage drop across a few turns of the inductor coil.

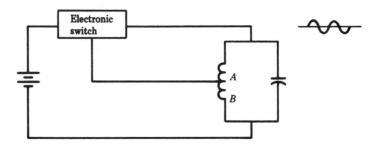

The small voltage across the few turns, between point B at the end of the coil
and point A, here about halfway along the coil, is used to operate the electronic switch.
       The use of a small part of an output voltage in this manner is called
feedback since it is "fed back" to an earlier part of the circuit to make it
operate correctly.
       When such a circuit is properly set up it will produce a continuous sine
wave output of constant amplitude and constant frequency. This circuit is
called an *oscillator.* The frequency of the sine waves generated by an oscillator is given by the following.

$$f = \frac{1}{2\pi\sqrt{LC}}$$

Practical oscillator circuits are shown in Chapter 9, but they all utilize the
principle shown here.

SELF-TEST

The questions below will test your understanding of this chapter. Use a separate sheet of paper for your diagrams or calculations. Compare your answers with the answer provided following the test.

1. What is the formula for the impedance of a series LC circuit?

    _____

2. What is the formula for the impedance of a series RLC circuit (once again, a circuit containing resistance, inductance, and capacitance)?

    _____

3. What is the relationship between X_C and X_L at the resonant frequency?

    _____

4. What is the voltage across the resistor in a series RLC circuit at the resonant frequency? _____

5. What is the voltage across a resistor in series with a parallel LC circuit at the resonant frequency? _____

6. What is the impedance of a series circuit at resonance? _____

7. What is the formula for the impedance of a parallel circuit at resonance?

    _____

8. What is the formula for the resonant frequency? _____

9. What is the formula for the bandwidth of a circuit? _____

10. What is the formula for the Q of a circuit? _____

In questions 11–13 a series LC circuit is used. In each case the values of the L, C, and R are given. Find f_r, X_L, X_C, Z, Q, and BW. Draw a rough curve of the output.

11. $L = 0.1$ mH, $C = 0.01$ μF, $R = 10$ ohms

    _____

    _____

12. $L = 4$ mH, $C = 6.4$ μF, $R = 0.25$ ohms

_____

_____

13. $L = 16$ mH, $C = 10$ μF, $R = 20$ ohms

_____

_____

In questions 14 and 15 a parallel circuit is used. No R is used; r is given. Find the same items as in questions 11–13.

14. $L = 6.4$ mH, $C = 10$ μF, $r = 8$ ohms

_____

_____

15. $L = 0.7$ mH, $C = 0.04$ μF, $r = 1.3$ ohms

_____

_____

16. Look at the following graph, which was taken for the output of a resonant circuit, and answer the questions.

(a) What is the peak value of the output curve? _____

(b) What is the resonant frequency? _____

(c) What is the voltage level at the half power points? _____

_____

(d) What are the half power frequencies? _____

(e) What is the bandwidth? _____

(f) What is the Q of the circuit? _____

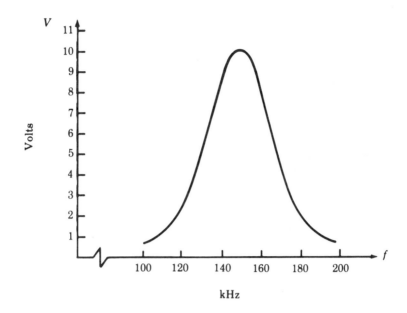

Answers to Self-Test

If your answers do not agree with those given below, review the frames indicated in parentheses before you go on to the next chapter.

1. $Z = X_L - X_C$ (frame 2)

2. $Z = \sqrt{(X_L - X_C)^2 + R^2}$ (frame 2)

3. $X_L = X_C$ (frame 5)

4. maximum output   (frame 5)

5. minimum output   (frame 11)

6. Z = minimum. Ideally it is equal to the resistance.   (frame 5)

7. $Z = \dfrac{L}{Cr}$, where r is the resistance of the coil   (frame 10)

8. $f_r = \dfrac{1}{2\pi\sqrt{LC}}$ (frame 6)

9. $\text{BW} = \dfrac{f_r}{Q}$ (frame 20)

10. $Q = \dfrac{X_L}{R}$ or $\dfrac{X_L}{r}$ (frame 20)

For your rough curve on questions 11-13, use the graph in frame 27 as a guide and insert the appropriate bandwidth and frequency values.

        (frames 21-29)

11. f_r = 160 kHz, X_L = X_C = 100 ohms, Q = 10, BW = 16 kHz, Z = 10 ohms

12. f_r = 1 kHz, X_L = 25 ohms, Q = 100, BW = 10 Hz, Z = 0.25 ohms

13. f_r = 400 Hz, X_L = 40 ohms, Q = 2, BW = 200 Hz, Z = 20 ohms

14. Since Q is not given, we should use the more complicated formula for finding the resonant frequency. This yields a value of about 600 Hz. X_L = 24 ohms and X_C = 26.5 ohms. Note they do not have to be equal when in a parallel circuit, but they are approximately equal when Q > 10. Z = 80 ohms and Q = 3 (small Q). The bandwidth is 200 Hz.

15. Again using the more complicated formula, we get f_r = 30 kHz which is the same result we would get when using the simpler formula. This tells us that X_L and X_C will be about equal and Q will be greater than 10. We get about 132 ohms for each reactance. Q = 101.5 and Z = 13.4 kΩ. The bandwidth is approximately 300 Hz.

16. (a) 10.1 V
    (b) 148 kHz
    (c) 10.1 × 0.707 = 7.14 V
    (d) approximately 135 kHz and 160 kHz (not quite symmetrical)
    (e) BW = 25 kHz
    (f) $Q = \dfrac{f_r}{BW}$ = about 5.9

CHAPTER EIGHT

Transistor Amplifiers

Many of the AC signals we deal with are very small. For example, the output from a record pick-up cannot drive a speaker, and the signal from a microphone cannot directly become part of the radio signal we receive in our homes. In each case an amplifier is needed.

The basics of amplifying a minute signal to a usable level can best be demonstrated by using a one-transistor amplifier, since this is the starting point for many modern electronic amplifying devices such as hi-fis, televisions, and radios.

Many amplifier circuit configurations are possible; the imagination is almost the only limit. The simplest and most basic of the possible amplifying circuits are presented here, in order to demonstrate how a transistor amplifies a signal and to give an easy procedure for designing an amplifier.

The emphasis in this chapter will be on the BJT just as it was in Chapters 3 and 4, which dealt primarily with the application of transistors in switching circuits. We will, however, demonstrate some other types of devices used as amplifiers. In particular, this includes the JFET (already introduced in Chapters 3 and 4) and an integrated circuit called the *op-amp* (operational amplifier).

When you complete this chapter you will be able to:

- calculate the voltage gain for an amplifier;

- calculate the DC output voltage for an amplifier circuit;

- select the appropriate resistors to design a circuit and provide the required gain;

- identify several ways of increasing the gain of a one-transistor amplifier;

- distinguish between the effects of a standard one-transistor amplifier and the emitter follower circuit;

- design a simple emitter follower;

- analyze a simple circuit to find the DC level out and the AC gain;
- design a simple common source (JFET) amplifier;
- analyze a JFET amplifier to find the AC gain;
- recognize an op-amp and its connections;
- apply the op-amp as an amplifier.

1. In Chapter 3 you learned how to turn transistors ON and OFF. You also learned how to make the output or collector DC voltage equal to half the power supply voltage. To review this concept, examine the circuit shown here.

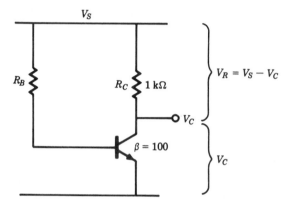

Since the power supply voltage $V_S = 10$ V, we want to find the value of R_B which will set the collector DC voltage V_C to be 5 V. The steps for determining R_B are shown below.

(1) Find I_C. Use $I_C = \dfrac{V_R}{R_C} = \dfrac{V_S - V_C}{R_C}$.

(2) Find I_B. Use $I_B = \dfrac{I_C}{\beta}$.

(3) Find R_B. Use $R_B = \dfrac{V_S}{I_B}$.

For the circuit above, use these values and the steps to find the value of R_B which will set the collector output voltage to 5 V.

$V_S = 10$ V, $R_C = 1$ kΩ, $\beta = 100$

(a) $I_C = $ _____

(b) $I_B = $ _____

(c) $R_B = $ _____

- - - - - - - - - - - - - - -

(a) $I_C = \dfrac{5\text{ V}}{1\text{ k}\Omega} = 5\text{ mA}$

(b) $I_B = \dfrac{5}{100} = 0.05\text{ mA}$

(c) $R_B = \dfrac{10\text{ V}}{0.05\text{ mA}} = 200\text{ k}\Omega$

2. You have discovered that using a 200 kΩ resistor for R_B will give an output level of 5 V at the collector. This procedure of setting the output DC level is called *biasing*. We have just biased the transistor to a 5 V DC output.

Examine the circuit and your calculations again.

(a) If R_B is decreased, what happens to I_B, I_C, V_R, and the bias point V_C? _____

_____

(b) If R_B is increased, what happens to the above measurements?

_____

– – – – – – – – – – – – – – – –

(a) I_B will increase, I_C will increase, V_R will increase, and so the bias point V_C will drop towards ground.

(b) They will all be the opposite of the above.

3. By making slightly different choices of R_B, we can produce slightly different values of I_B.

The slight variations in I_B will be amplified by the transistor by an amount equal to the current gain, β, of the transistor, and these amplified variations will appear in the collector current.

The variations in collector current will cause variations in the voltage drop V_R across R_C and so the output voltage measured at the collector will also vary.

Using $R_B = 168$ kΩ in the same circuit, find the following.

(a) $I_B = \dfrac{V_S}{R_B} =$ _____

(b) $I_C = \beta I_B =$ _____

(c) $V_R = I_C R_C =$ _____

(d) $V_C = V_S - V_R =$ _____

– – – – – – – – – – – – – – – –

(a) $I_B = \dfrac{10 \text{ V}}{168 \text{ k}\Omega} = 0.059 \text{ mA}$

(b) $I_C = 100 \times 0.059 = 5.9 \text{ mA}$

(c) $V_R = 1 \text{ k}\Omega \times 5.9 \text{ mA} = 5.9 \text{ V}$

(d) $V_C = 10 \text{ V} - 5.9 \text{ V} = 4.1 \text{ V}$

4. Now use these values of R_B and find the corresponding values of V_C.

(a) 100 kΩ _____

(a) 10 MΩ _____

(c) 133 kΩ _____

(d) 400 kΩ _____

- - - - - - - - - - - - - - - - - -

(a) $I_B = 0.1 \text{ mA}$, $I_C = 10 \text{ mA}$, $V_C = 0 \text{ V}$

(b) $I_B = 1 \text{ } \mu\text{A}$, $I_C = 0.1 \text{ mA}$, $V_C = 10 \text{ V}$ (approximately)

(c) $I_B = 0.075 \text{ mA}$, $I_C = 7.5 \text{ mA}$, $V_C = 2.5 \text{ V}$

(d) $I_B = 0.025 \text{ mA}$, $I_C = 2.5 \text{ mA}$, $V_C = 7.5 \text{ V}$

5. The axes for I_C and V_C are shown in the graphs below. When the points or values found in frame 4 are plotted and joined, we get a straight line, called the *load line*.

   The axis labeled V_C is actually the voltage measured between the collector and the emitter of the transistor, and not that measured between the collector and ground. Thus it should be labeled V_{CE}.

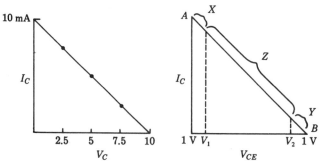

(a) At point A in the graph on the right, is the transistor ON or OFF?

_____

(b) Is it ON or OFF at point B? _____

- - - - - - - - - - - - - - - -

(a) ON because full current flows, and the transistor acts like a short circuit. The voltage drop across the transistor is 0 V.

(b) OFF because essentially no current flows, and the transistor acts like an open circuit. The voltage drop across the transistor will be at its maximum, 10 V in this case.

6. In the graph in frame 5, point A is often called the *saturated point* or the *saturation point*, since the collector current is at a maximum and no more can flow.

Point B is often called the *cut off* point since no collector current flows because it is cut off.

In regions X and Y the ratio of $\beta = I_C/I_B$ is not constant, so these are called the *nonlinear* regions. The reason why has to do with the physical construction of the transistor and we will not discuss it here.

As a rough guide, V_1 is about 1 V, and V_2 is about 1 V less than the voltage at point B.

What kind of voltage is indicated at point B? _____

— — — — — — — — — — — — — — — —

the DC supply voltage

7. Region Z in the graph in frame 5 is called the *linear region*, where β is constant. During amplification the voltage and current swings of the output signal should all occur in the linear region. This will ensure the output will be a faithfully amplified copy of the input. In other words, it will not be distorted.

Indicate which set of values below would give an undistorted amplified output.

_____ (a) $I_C = 9$ mA, $V_C = 1$ V

_____ (b) $I_C = 1$ mA, $V_C = 9$ V

_____ (c) $I_C = 6$ mA, $V_C = 4.5$ V

— — — — — — — — — — — — — — — —

c is the only one. a and b fall into the nonlinear regions.

8. If a small AC signal is applied to the base of the transistor after it has been biased, the small voltage variations will cause small variations in the base current.

These will be amplified by a factor of β and will cause corresponding variations in the collector current. These in turn will cause similar variations in the collector voltage.

It should be pointed out that the β used for AC gain calculations is different from the β used in calculating DC variations. The AC β is the value of the common emitter, AC forward current transfer ratio, when the transistor is being operated properly. It is listed as h_{fe} in manufacturers' data sheets for individual transistors. The AC β is used whenever you need to calculate the AC output for a given AC input or determine an AC current variation. The DC β that has been used up to now was used only in the context of the relationship between the base and the collector DC current values. It is important to know which β to use and to remember that one is for DC and the other is for AC variations. The DC β is sometimes called h_{FE} or β_{dc}.

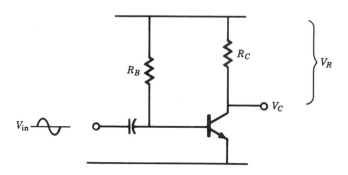

As V_{in} increases it adds to the base current, and so causes the base current to increase. This causes the collector current to increase, thus causing the voltage drop across R_C to increase. So V_C decreases.

The capacitor shown at the input will block DC (infinite reactance) and easily pass AC (low reactance). This is a common isolation technique at AC inputs and outputs and will be seen often in this book.

(a)   If the input signal decreases, what will happen to the collector

voltage? _____

_____

(b)   If a sine wave is applied to the input, what shape would you expect

at the collector? _____

_____

— — — — — — — — — — — — — — —

(a)   The collector voltage, V_C, will increase.
(b)   a sine wave, but inverted as in the diagram in the next frame.

9. Look at this circuit and the input and output voltage waveforms.

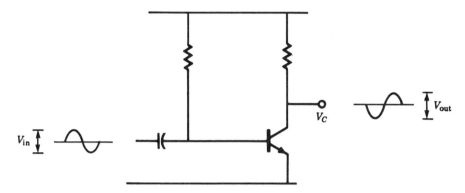

The input voltage V_{in} is applied to the base. (Strictly it is applied between the base and the emitter). Now look at the voltage variations at the collector. These are centered around the DC bias point V_C, and they will be larger than the input voltage variations; that is, they are amplified.

These amplified variations at the collector will later be used to drive some components, such as a speaker, and are called the *output voltage* or *output signal*.

To distinguish these AC output variations from the DC bias level, the AC output voltage is indicated by V_{out}. In most cases this will refer to the peak-to-peak value, unless otherwise stated.

(a) What is meant by V_C? _____

(b) What is meant by V_{out}? _____

— — — — — — — — — — — — — — — — —

(a) collector DC voltage, or the bias point
(b) AC output voltage

The ratio of the output voltage to the input voltage is called the *voltage gain* of the amplifier.

$$\text{Voltage gain} = A_V = \frac{V_{out}}{V_{in}}$$

This voltage gain can be found directly by measuring the AC voltage in and out with an oscilloscope. When doing this, look at the AC peak-to-peak voltages only. Disregard the DC levels. Gain is an AC concept, it has nothing to do with the DC bias point at all.

10. In the simple circuit we have been using up to now, the voltage gain can be calculated using the following formula.

$$A_V = \beta \times \frac{R_L}{R_{in}}$$

where:

R_L = is the load resistance. In this circuit it is the collector resistor R_C only.

R_{in} = the *input resistance* or *input impedance* of the transistor. R_{in} can be found on the manufacturers' data or specification sheets and is often called h_{ie}. In most transistors it is about 1 kΩ to 2 kΩ.

In the circuit for frame 9, assume the input impedance is 1 kΩ, V_{in} = 1 mV, R_C = 1 kΩ, and β = 100. We want to find V_{out} by combining the two voltage gain equations.

$$A_V = \frac{V_{out}}{V_{in}} \quad \text{and} \quad A_V = \beta \times \frac{R_L}{R_{in}}$$

$$\frac{V_{out}}{V_{in}} = \beta \times \frac{R_L}{R_{in}}$$

Thus:

$$V_{out} = V_{in} \times \beta \times \frac{R_L}{R_{in}}$$

$$= 1 \text{ mV} \times 100 \times \frac{1 \text{ k}\Omega}{1 \text{ k}\Omega}$$

$$= 100 \text{ mV}$$

(a) Now repeat this using an input impedance of 2 kΩ. _____

(b) Find the voltage gain in both cases. _____

– – – – – – – – – – – – – – – –

(a) V_{out} = 50 mV
(b) A_V = 100 and A_V = 50

This simple amplifier can provide voltage gains up to about 500. But it does have several faults which limit its practical usefulness.

(1) Due to variations in β between transistors, V_C will change if the transistor is changed. To correct this R_B will have to be changed or made variable.

(2) R_{in} or h_{ie} varies greatly from transistor to transistor. This variation, combined with the variations in β means that the gain cannot be guaranteed from one transistor amplifier to another.

(3) Both R_{in} and β change greatly with temperature, hence the gain is very temperature dependent. For example, a simple amplifier circuit like that shown here, which was designed to work in the desert in July would fail completely in Alaska in winter. If it worked perfectly in the lab, it probably would not work outdoors on either a hot or cold day.

An amplifier whose gain and DC level bias point change as described above is said to be "unstable." For reliable operation an amplifier should be as stable as possible. In later frames you will see how a stable amplifier is designed.

Optional Experiment

If you have the appropriate facilities you will find it helpful to perform the experiment described here. The following equipment is required.

a transistor
250 kΩ potentiometer (i.e., a variable resistor)
1 kΩ resistor
10 kΩ resistor
0.1 μF capacitor
9 V transistor radio battery or a lab type power supply
a signal generator or a sine wave oscillator
an oscilloscope

You may possibly also need several "clip leads" to connect the components together to make the circuit. Set up the following circuit as shown.

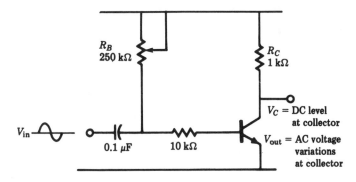

Use this procedure to perform the experiment.

(1) Adjust the potentiometer so that the collector DC level—the bias point—is 4.5 V to 5 V.
(2) Connect the signal generator to the input and set it to give a sine wave output of 1 kHz.
(3) Adjust the level of the signal generator so that the output of the

transistor viewed at the collector is not distorted. Make sure it looks like a sine wave and not like this.

(4) Measure the pp sine wave input and output of the transistor.
(5) Find the voltage gain from the formula $A_V = V_{out}/V_{in}$.
(6) Now, heat the transistor while it is in the circuit. Do this by placing a soldering iron near it for 15–30 seconds. Note the changes in the DC level at the collector and the changes in the pp level of the output sine wave.
(7) Repeat steps 1–6 with other transistors. Note the differences in gain.

You will undoubtedly have discovered that you cannot always guarantee which way the outputs will change. This is because the relative values of β and R_{in} may be different and they may change at different rates as the temperature changes.

The point of this experiment is twofold.

(1) To demonstrate the voltage gain of a simple transistor amplifier.
(2) To show how such a simple amplifier is "unstable." It is very important to notice that the bias point and the gain can change, and in fact will change. Obviously this limits the practical usefulness of this circuit. To design a usable stable amplifier a few changes are required. These are covered in the next few frames.

This is a good stopping point, if you want to take a break.

A STABLE AMPLIFIER

11. To overcome the instability of the previous simple circuit, three changes are made to the basic amplifier circuit. Two resistors, R_1 and R_2, are added, to ensure the stability of the DC bias point.

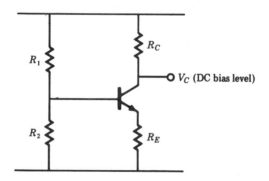

The addition of the emitter resistor R_E ensures the stability of the AC gain.

The circuit is redrawn below with the important DC currents and voltage shown. These will be used in the remainder of this chapter.

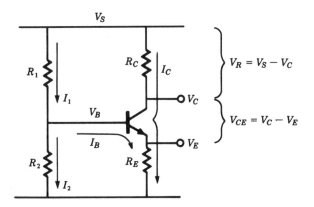

In designing an amplifier circuit and choosing the resistor values, we are trying to achieve two objects. What are they? _____

_____

– – – – – – – – – – – – – – –

a stable DC bias point, and a stable AC gain

12. Let's look at the gain first. The gain formula, once again, is as follows.

$$A_V = \frac{V_{out}}{V_{in}} = \frac{R_C}{R_E}$$

This is a slight variation on the formula shown in frame 10. (The complex mathematical justification for this is not covered here.) The AC gain now has been made independent of the transistor β and the input impedance.

As these two parameters vary with temperature, and vary from transistor to transistor, we now have a method of setting the AC gain which will be constant regardless of all these variations.

Suppose a circuit has $R_C = 10$ kΩ and $R_E = 1$ kΩ.

(a)   What is the AC voltage gain for a transistor with a $\beta = 100$? _____

(b)   What is the gain when $\beta = 500$? _____

– – – – – – – – – – – – – – –

(a)   10
(b)   10

13. A couple of numerical examples will help fix the relationships in your mind.

    (a) If $R_C = 10$ kΩ and $R_E = 1$ kΩ, what will be the voltage gain A_V of the amplifier? If 2 mV$_{pp}$ is the input signal, what is the pp output voltage? _____

    _____

    (b) If $R_C = 1$ kΩ and $R_E = 250$ ohms, what is the voltage gain? What is the output voltage if the input is 1 V$_{pp}$? _____

    _____

- - - - - - - - - - - - -

    (a) $A_V = \dfrac{R_C}{R_E} = \dfrac{10 \text{ k}\Omega}{1 \text{ k}\Omega} = 10$

    $V_{out} = 10 \times V_{in} = 20$ mV

    (b) $A_V = \dfrac{1 \text{ k}\Omega}{250 \text{ ohms}} = 4$

    $V_{out} = 4 \text{ V}_{pp}$

    Although the amplifier circuit will produce almost exact values for the gain, it does not produce high values of gain. For various reasons this circuit is limited to gains of 50 or less. For larger gains, another circuit variation is shown later.

14. Before we continue, let's look at the current relationships in this simple amplifier and explain a common approximation which is often made. Consider the emitter current in frame 11.

    $I_E = I_B + I_C$

    In other words, it is the sum of the base and the collector currents.
        Now, in most transistors the collector current is much larger than the base current as β is a very large number.

    $I_C = \beta I_B$

    The common usage, then, is to regard the emitter and collector currents to be the same, disregarding the base current. Thus, $I_C = I_E$.
        In the circuit on the next page, find V_C, V_E, and A_V. Also find the voltage between the collector and the emitter.

    $V_S = 10$ V, $I_C = 1$ mA. Use $R_C = 1$ kΩ and $R_E = 100$ ohms.

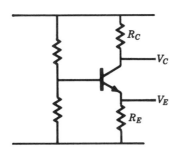

R_C
R_E

- - - - - - - - - - - - - - - - - -

R_C R_E

$V_R = 1\ \text{k}\Omega \times 1\ \text{mA} = 1\ \text{V}$

$V_C = V_S - V_R = 10 - 1 = 9\ \text{V}$

$V_E = 100\ \text{ohms} \times 1\ \text{mA} = 0.1\ \text{V}$

$A_V = \dfrac{V_R}{V_E} = \dfrac{1}{0.1} = 10$

$A_V = \dfrac{R_C}{R_E} = \dfrac{1\ \text{k}\Omega}{100\ \text{ohms}} = 10$

15. Use the same circuit with $R_C = 2\ \text{k}\Omega$ and $R_E = 1\ \text{k}\Omega$. Again find V_C, V_E, and A_V.

- - - - - - - - - - - - - - - - - -

$V_R = 2\ \text{k}\Omega \times 1\ \text{mA} = 2\ \text{V}$

$V_C = 10 - 2 = 8\ \text{V}$

$V_E = 1\ \text{k}\Omega \times 1\ \text{mA} = 1\ \text{V}$

$A_V = \dfrac{V_R}{V_E} = \dfrac{2\ \text{V}}{1\ \text{V}} = 2$

$A_V = \dfrac{R_C}{R_E} = \dfrac{2\ \text{k}\Omega}{1\ \text{k}\Omega} = 2$

16. For the same circuit, find V_C, V_E, and A_V for these sets of values.

(a) $R_C = 5\ \text{k}\Omega$, $R_E = 1\ \text{k}\Omega$

- - - - - - - - - - - - -

(b) $R_C = 4.7\ \text{k}\Omega$, $R_E = 220\ \text{ohms}$

- - - - - - - - - - - - - - - -

(a) $V_R = 5\ \text{V}$, $V_C = 5\ \text{V}$, $V_E = 1\ \text{V}$, $V_{CE} = 4\ \text{V}$, $A_V = 5$

(b) $V_R = 4.7\ \text{V}$, $V_C = 5.3\ \text{V}$, $V_E = 0.22\ \text{V}$, $V_{CE} = 5.08\ \text{V}$, $A_V = 21.36$

BIASING

17. We will now see how to bias our circuit, or set the DC output voltage level.

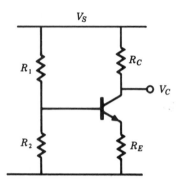

We have to find values for R_1, R_2, and R_E which will bias the circuit to its correct DC conditions and provide the required gain.

In order to start the bias design, three factors must be stated as desired conditions.

The AC gain required.
The DC output voltage level.
The value of the collector resistor. (We will not go into the reasons for this third factor.)

Read the following procedure and the relevant formulas, then we will walk through an example.

(1) Find R_E. Use $A_V = \dfrac{R_C}{R_E}$.

(2) Find V_E. Use $A_V = \dfrac{V_R}{V_E} = \dfrac{V_S - V_C}{V_E}$.

(3) Find V_B. Use $V_B = V_E + 0.7$ V.

(4) Find I_C. Use $I_C = \dfrac{V_S - V_C}{R_C}$.

(5) Find I_B. Use $I_B = \dfrac{I_C}{\beta}$.

(6) Find I_2. Refer to the circuit. Choose I_2 to be $10I_B$. (This can be justified mathematically as the crucial step in providing the stability of the DC bias point, but the details are beyond the scope of this book.)

(7) Find R_2. Use $R_2 = \dfrac{V_B}{I_2}$.

(8) Find R_1. Use $R_1 = \dfrac{V_S - V_B}{I_2 + I_B}$.

(9) Usually steps 7 and 8 produce nonstandard values for the resistors. So now choose the nearest standard values.

(10) Use the voltage divider formula to see if the standard values chosen in step 9 give a voltage level close to V_B found in step 3. ("Close" means within 10% of the ideal.)

This procedure will produce an amplifier which will work, and with DC voltage levels and AC gain value close to that originally desired.

Now apply this procedure to an example. Assume the given values are the following.

$$A_V = 10, \ V_C = 5 \text{ V}, \ R_C = 1 \text{ k}\Omega, \ \beta = 100$$

Find the values of the other resistors in the circuit shown below.

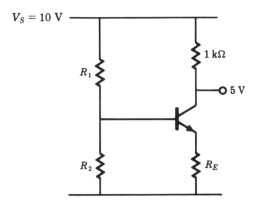

$V_S = 10$ V

R_1

R_2

$1 \text{ k}\Omega$

5 V

R_E

Work through the steps 1 through 10, referring back for formulas if necessary.

(1) Find R_E. $A_V = \dfrac{R_C}{R_E}$. So $R_E = \dfrac{R_C}{A_V} = \dfrac{1 \text{ k}\Omega}{10} = 100$ ohms.

(2) Find V_E. $V_E = $ _____.

(3) Find V_B. $V_B = $ _____.

(4) Find I_C. $I_C = $ _____.

(5) Find I_B. $I_B = $ _____.

(6) Find I_2. $I_2 = $ _____.

(7) Find R_2. $R_2 = $ _____.

(8) Find R_1. $R_1 = $ _____.

(9) Choose the nearest standard values.

$R_1 = $ _____ ; $R_2 = $ _____.

(10) Using the values for R_1 and R_2, find V_B.

$$V_B = \underline{\hspace{8cm}}.$$

You should have produced values close to these.

(1)   100 ohms
(2)   0.5 V
(3)   1.2 V
(4)   5 mA
(5)   0.05 mA
(6)   0.5 mA
(7)   2.4 kΩ
(8)   16 kΩ
(9)   2.4 kΩ and 16 kΩ are standard values (they are 5% values). These can be used directly. Alternative acceptable values would be 2.2 kΩ and 15 kΩ.
(10) With 2.4 kΩ and 16 kΩ, $V_B = 1.3$ V. With 2.2 kΩ and 15 kΩ, $V_B = 1.28$ V. Either value of V_B is acceptable in practice as being close to the value of 1.2 V in step 3.

Optional Experiment

Set up the following circuit, just as you did the one earlier in this chapter. Once it is set up apply the DC power supply voltage only; do not supply the AC input voltage yet. Measure the DC voltages at the base, emitter, and collector, and see how closely they compare to the calculated values.

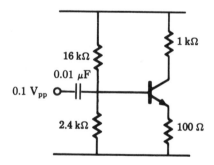

Now apply the signal generator and provide an AC sine wave input of about 0.1 V_{pp}. Use a frequency of 1 kHz. Use the oscilloscope to measure the pp voltages at the base, emitter, and the collector. Verify that the emitter and base voltages are the same, and that the collector pp voltage is about 10 times the input voltage.

Leave this setup as you read the next frame. If you did not perform this experiment, draw the circuit on a separate sheet of paper and label it for reference.

18. Here is an example for you to work, in which you will design a simple amplifier which will meet the specifications given.

With V_S = 10 V and R_C = 3.3 kΩ, bias the amplifier to have an output voltage of 6 V and a gain of 15. Assume β = 100.

Just follow the steps given in frame 17, and check your answers with those given below as you work them out.

(1)   Find R_E. R_E = _____.

(2)   Find V_E. V_E = _____.

(3)   Find V_B. V_B = _____.

(4)   Find I_C. I_C = _____.

(5)   Find I_B. I_B = _____.

(6)   Find I_2. I_2 = _____.

(7)   Find R_2. R_2 = _____.

(8)   Find R_1. R_1 = _____.

(9)   Nearest standard values.

   R_1 = _____; R_2 = _____.

(10) Check V_B. V_B = _____.

— — — — — — — — — — — — — — — — —

These are the figures you should have obtained.

(1)   220 ohms
(2)   0.27 V
(3)   0.97 V (You can use 1 V if you wish.)
(4)   1.2 mA
(5)   0.012 mA
(6)   0.12 mA
(7)   8.3 kΩ
(8)   68.2 kΩ
(9)   These are very close to the standard values of 8.2 kΩ and 68 kΩ.
(10) 1.08 using the standard values, close enough

19. In the circuit we have just designed the AC voltage gain was 10. Earlier we pointed out that the maximum practical gain with this circuit is about 50.

In the original circuit for frame 1, AC gains of up to 500 are possible. So by ensuring the stability of the DC bias point, we have lost the high gain possible with the transistor.

Is it possible to have a high AC gain from an amplifier with stable bias points? Yes, it is possible. This can be achieved by placing a capacitor in parallel with the emitter resistor, as shown on page 190.

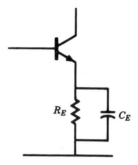

This capacitor effectively grounds the emitter as far as the AC signal is concerned. So the AC signal "sees" a different circuit from the DC. The two circuits are shown below.

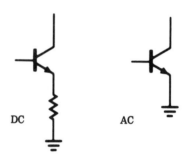

The AC gain is now very close to that of the original amplifier circuit shown in frame 1.

(a) What effect does the capacitor have on an AC signal? _____

_____

(b) What effect does it have on the AC gain? _____

_____

(c) What is the AC gain formula now? _____

- - - - - - - - - - - - - - - -

(a) It makes the emitter look like a ground, and effectively turns the circuit into the original one we used.

(b) It increases the gain.

(c) $A_V = \beta \times \dfrac{R_C}{R_{in}}$

20. The circuit of frame 19 is used when as much AC gain as possible is required. Predicting the actual amount of gain is usually not important in this situation, and the fact that the equation is inexact does not

matter. If an accurate amount of gain is required, then lower amounts of gain must be accepted.

The value of the capacitor C_E is found using the procedure which follows.

(1)   Decide the lowest frequency the amplifier will have to handle.

(2)   Choose X_C to be $\dfrac{R_E}{10}$ at this lowest frequency.

(3)   Calculate C_E from $X_C = \dfrac{1}{2\pi fC}$.

In the circuit you constructed or drew earlier, we wish to pass a reasonably low audio frequency of 50 Hz. Find the value of C_E required. The emitter circuit is shown below.

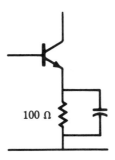

$100 \ \Omega$

(1)   50 Hz has been chosen as the lowest frequency.

(2)   Choose X_C. $X_C =$ _____.

(3)   Calculate C_E. $C_E =$ _____.

- - - - - - - - - - - - - - - - -

$X_C = 10$ ohms
$C_E = 320 \ \mu F$ (approximately)

Are the two gain formulas for the amplifier really the same? The actual gain formula for the amplifier is as follows.

$$A_V = \beta \times \dfrac{R_C}{R_{in} + R_E}$$

(We are using R_C instead of R_L, since the collector resistor is the total load on the amplifier.)

We have three cases to consider, shown by these three figures.

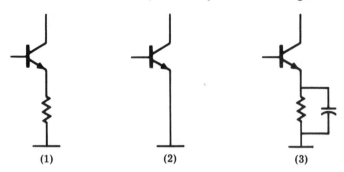

(1)   (2)   (3)

Case 1: Here βR_E is much greater than R_{in}, and the formula simplifies to:

$$A_V = \beta \times \frac{R_C}{\beta R_E} = \frac{R_C}{R_E}.$$

Case 2: Here $R_E = $ zero, so the formula becomes:

$$A_V = \beta \times \frac{R_C}{R_{in}}.$$

Case 3: Here $R_E = $ zero, since the AC signal is grounded by the capacitor and R_E is out of the AC circuit. Thus the formula becomes:

$$A_V = \beta \times \frac{R_C}{R_{in}}.$$

21. To obtain even larger voltage gains, two transistor amplifiers can be *cascaded*. That is, the output of the first can be fed into the input of the second. The circuit below shows a two-transistor amplifier made by cascading.

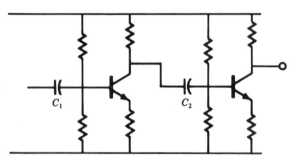

As shown, the amplifiers are designed individually and are connected by capacitor C_2. The particular capacitor is chosen so that it has a very low reactance at the lowest frequency required to be passed.

The overall gain is then found by multiplying the individual gains. For example, if the first amplifier has a gain of 10 and the second has a gain of 10, then the overall gain is 100.

(a) Suppose an amplifier with a gain of 15 is cascaded with one that has a gain of 25. What is the overall gain? _____

(b) What is the overall gain if the individual gains are 13 and 17?

_____

_ _ _ _ _ _ _ _ _ _ _ _ _ _ _ _

(a) 375
(b) 221

22. If capacitors are placed in parallel with each emitter resistor, very large gains can be achieved.

For example, each "stage" of the two-transistor amplifier may have an individual gain of 100. What is the overall gain? _____

_ _ _ _ _ _ _ _ _ _ _ _ _ _ _

10,000

Gains of several thousand are not uncommon in electronics. Suppose we want a very large gain, but wish to be able to have some specified value of gain.

A way of achieving this is to use a *bypassed* emitter, to obtain maximum gain, and to adjust the level of the output voltage with a variable resistor. Note the two methods, shown in the following diagram, of doing this. Method 2 is the most common, and is often found as the "volume control" in consumer equipment.

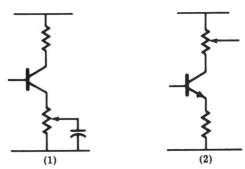

(1)   (2)

THE EMITTER FOLLOWER

23. A variation on the amplifier is the circuit shown here.

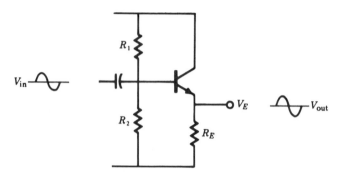

How is this different from the earlier amplifier circuit? _____

_____

— — — — — — — — — — — — — — — — —

There is no collector resistor, and the output signal is taken from the
emitter.

24. The circuit in frame 23 is called an *emitter follower.* (In some books it
is also called the *common collector* amplifier.)
     The output signal of the emitter has some interesting features.

Its peak-to-peak value is almost the same as the input signal. In
other words, the circuit gain is slightly less than 1. In practice it is
often considered to be 1.

The output is the same phase as the input. It is not inverted. In
fact, the output is simply considered to be the same as the input.

It has a very high input resistance, and so it draws very little current from the signal source.

It has a very low output resistance. So the signal at the emitter
appears to be emanating from a battery or signal generator with
a very low internal resistance.

(a) What is the voltage gain of an emitter follower? _____

(b) Is the output inverted or not? _____

(c) What is the input resistance to the emitter follower? _____

(d) What is its output resistance? _____

— — — — — — — — — — — — — — —

(a)   1
(b)   not inverted
(c)   high
(d)   low

25. This example will demonstrate the importance of the emitter follower
     circuit. Imagine a small AC motor with 100 ohms resistance which must
     be driven by a 10 V_{pp} signal from a generator. But it is discovered the
     generator has an internal resistance of 50 ohms.
         From the circuit shown below, it is easy to see that the voltage
     across the motor is only about 6.7 V_{pp}.

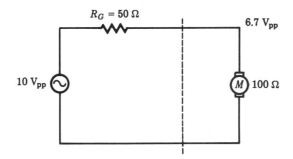

Now we will interpose an emitter follower between the generator and
the motor.

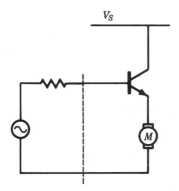

         The input resistance of the emitter follower is approximated by
this formula.

$$R_{in} = \beta \times R_E$$

So the load on the generator is now provided by the emitter follower
instead of the motor, and is now 10,000 ohms instead of 100 ohms.
This will not cause the generator voltage to drop, and so it remains at
10 V_{pp}. The emitter voltage also remains at 10 V_{pp}.

But the current through the motor is now produced by the power supply and not the generator, and the transistor looks like a generator with a very low internal resistance.

This internal resistance is called the output impedance of the emitter follower. It can be approximated by the formula below.

$$R_O = \frac{\text{internal resistance of generator}}{\beta}$$

Here the internal resistance is about 0.5 ohms.

So now we effectively have a load of 100 ohms connected to a generator with an internal resistance of only 0.5 ohms. Hence the output voltage of 10 V_{pp} will be maintained across the motor.

(a)   What was the purpose of the emitter follower in the above

example? _____

_____

(b)   Which two properties of the emitter follower are important for

use in circuits? _____

_____

– – – – – – – – – – – – – – – –

(a)   to drive a load which could not be driven directly or directly connected to a generator
(b)   high input resistance and its low output resistance

26.   Before we look at how to bias an emitter follower, try to answer these questions.

(a)   What is the approximate gain of the circuit? _____

(b)   What is the phase of the output signal compared to the input

signal? _____

(c)   Which has the higher value, the input resistance or the output

resistance? _____

(d)   Is the emitter follower more effective at amplifying signals or at

isolating loads? _____

– – – – – – – – – – – – – – – –

(a)   1
(b)   the same phase
(c)   the input resistance
(d)   isolating loads

27. An emitter follower circuit is designed similarly to the amplifier circuit shown in frame 17.

The voltage level at the emitter must be specified. The value of the emitter resistor is generally less than 1 kΩ. The steps below are used in sequence.

(1) Specify V_E. This is a DC voltage level, and usually half the supply voltage is chosen. This is always a good starting point.

(2) Find V_B. Use $V_B = V_E + 0.7$ V.

(3) Specify R_E. Often this is a given factor, especially if it is a motor or a meter which must be driven.

(4) Find I_E. Use $I_E = \dfrac{V_E}{R_E}$.

(5) Find I_B. Use $I_B = \dfrac{I_E}{\beta}$.

(6) Find I_2. Use $I_2 = 10I_B$.

(7) Find R_2. Use $R_2 = \dfrac{V_B}{I_2}$.

(8) Find R_1. Use $R_1 = \dfrac{V_S - V_B}{I_2 + I_B}$.

Usually here I_B is ignored.

(9) Choose the nearest standard values for R_1 and R_2.

(10) Check that these standard values will give a voltage close to V_B. Use the voltage divider formula.

A simple design example will illustrate this procedure. Use the values shown in the circuit below, and work through the procedure to find the values of the two bias resistors.

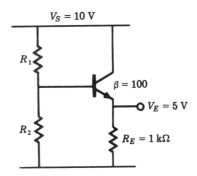

(1) $V_E = $ _____

(2) $V_B = $ _____

(3) $R_E = $ _____

(4) $I_E = $ _____

(5) $I_B =$ _____

(6) $I_2 =$ _____

(7) $R_2 =$ _____

(8) $R_1 =$ _____

(9)   The nearest standard values are:

$R_1 =$ _____ and $R_2 =$ _____

(10) $V_B =$ _____

— — — — — — — — — — — — — — —

Your answers should be close to these values.

(1)   5 V (This was already given.)
(2)   5.7 V
(3)   1 kΩ (This was a given value.)
(4)   5 mA
(5)   0.05 mA
(6)   0.5 mA
(7)   11.4 kΩ
(8)   7.8 kΩ
(9)   Nearest values are 8.2 kΩ and 12 kΩ.
(10) These will give 5.94 V. This is a little high, but is acceptable.

Because V_E is set by the biasing resistors, it is not dependent upon the value of R_E; almost any value of R_E can be used in this circuit. The minimum value for R_E is given by the simple equation below.

$$R_E = \frac{10 \, R_2}{\beta}$$

Because of the wide range of R_E the emitter follower circuit can be designed and implemented very quickly, and a search for exact bias resistor values is not necessary.

This is a good stopping point if you are ready to take a break.

ANALYZING AN AMPLIFIER

28.   Up to now the emphasis has been on designing a simple amplifier and an emitter follower. This section will show how to "analyze" a circuit which has already been designed. By "analyze" we mean calculate the collector DC voltage (the bias point) and find out the AC gain.

The procedure is basically the reverse of the design procedure. Let's start with this circuit.

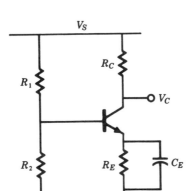

The procedure we will use is shown below.

(1) Find V_B. Use $V_B = V_S \times \dfrac{R_2}{R_1 + R_2}$.

(2) Find V_E. Use $V_E = V_B - 0.7$ V.

(3) Find I_C. Use $I_C = \dfrac{V_E}{R_E}$. Note that $I_C = I_E$.

(4) Find V_R. Use $V_R = R_C \times I_C$.

(5) Find V_C. Use $V_C = V_S - V_R$. This is the bias point.

(6) Find $A_V = \dfrac{R_C}{R_E}$ or $A_V = \beta \times \dfrac{R_C}{R_{in}}$. When using this second

   formula, the value of R_{in} (or h_{ie}) must be found from the
   manufacturer's data sheets for the transistor.

A simple example will help us here. Use the circuit shown below, with
the values given. Assume $\beta = 100$ and $R_{in} = 2$ kΩ.

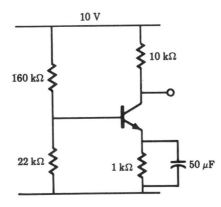

We are going to find V_B, V_E, I_C, V_R, V_C, and A_V. Use steps 1–6 as outlined above.

(1) $V_B =$ _____

(2) $V_E =$ _____

(3) $I_C =$ _____

(4) $V_R =$ _____

(5) $V_C =$ _____

(6) $A_V =$ _____

- - - - - - - - - - - - - - -

(1) $V_B = 10 \times \dfrac{22 \text{ k}\Omega}{160 \text{ k}\Omega + 22 \text{ k}\Omega} = 1.2$ V

(2) $V_E = 1.2 - 0.7 = 0.5$ V

(3) $I_C = \dfrac{0.5 \text{ V}}{1 \text{ k}\Omega} = 0.5$ mA

(4) $V_R = 10 \text{ k}\Omega \times 0.5 \text{ mA} = 5$ V

(5) $V_C = 10 \text{ V} - 5 \text{ V} = 5$ V (This is the bias point.)

(6) With the capacitor: $A_V = 100 \times \dfrac{10 \text{ k}\Omega}{2 \text{ k}\Omega} = 500$ (a large gain)

Without the capacitor: $A_V = \dfrac{10 \text{ k}\Omega}{1 \text{ k}\Omega} = 10$ (a small gain)

29. A further check we can make is the lowest frequency the amplifier will satisfactorily pass. For this, follow the simple procedure below.

  (1) Check the value of R_E.
  (2) Calculate the frequency at which $X_C = R_E/10$. Use the capacitor reactance formula. (This again is one of those "rules of thumb" which *can* be mathematically justified, and gives reasonably accurate results in practice.)

For the circuit in frame 28, find the following.

(a) $R_E =$ _____

(b) $f =$ _____

- - - - - - - - - - - - - - -

(a) $R_E = 1$ kΩ

(b) So we will set $X_C = 100$ ohms, and use this formula.

$$X_C = \frac{1}{2\pi fC}$$

$$100 \text{ ohms} = \frac{0.16}{f \times 50 \text{ }\mu\text{F}} \quad \text{since } 0.16 = \frac{1}{2\pi}$$

$$\text{so } f = \frac{0.16}{100 \times 50 \times 10^{-6}} = 32 \text{ Hz}$$

30. Now follow the same procedure to analyze the circuit below. As you get each step completed, check your figures with the answers.

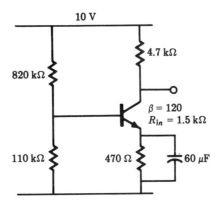

(1) V_B = _____

(2) V_E = _____

(3) I_C = _____

(4) V_R = _____

(5) V_C = _____

(6) With capacitor: A_V = _____

Without capacitor: A_V = _____

(7) Low frequency check: f = _____

— — — — — — — — — — — — — — — —

Your figures should be close to these.

(1) 1.18 V
(2) 0.48 V
(3) 1 mA
(4) 4.7 V
(5) 5.3 V (bias point)
(6) 376; 10
(7) 57 Hz (approximately)

THE JFET AS AN AMPLIFIER

31. Let us review the operation of the JFET as discussed in Chapter 3 (frames 29–32) and Chapter 4 (frames 37–41). You may wish to review these frames in answering the questions presented by this frame. The JFET shown is in a typical biasing circuit.

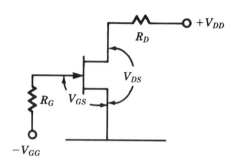

(a) What type of JFET is depicted in the circuit? _____

(b) What should the voltage from the gate to the source, V_{GS}, be in order to turn the JFET "hard ON"? _____

(c) What drain current will flow under the condition of (b)?

_____

(d) What should the voltage, V_{GS}, be in order to turn the JFET "hard OFF"? _____

(e) In changing V_{GS} between the two extremes in (b) and (d), what are we doing to the JFET? _____

- - - - - - - - - - - - - - - -

(a) N-channel JFET
(b) $V_{GS} = 0$ V to turn the JFET "hard ON"
(c) drain saturation current, I_{DSS}
(d) V_{GS} should be a large negative voltage for the N-channel JFET to turn it OFF. The voltage must be larger than or equal to the cutoff voltage.
(e) We are operating the JFET as a switch.

32. Now let's think in terms of operating the JFET with a gate to source voltage about halfway between the ON and OFF states. The drain current that would flow can be found from the following equation, which is a mathematical approximation of the transfer curve.

$$I_D = I_{DSS} \left(1 - \frac{V_{GS}}{V_P} \right)^2$$

where I_{DSS} is the value of the drain saturation current and V_P is the voltage at cutoff, $V_{GS(off)}$. These values can be seen as the endpoints of the transfer curve discussed in frame 38 of Chapter 4. A typical sketch of this curve for an N-channel JFET is repeated here.

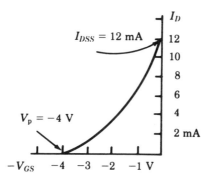

You can look at the curve and find the drain current for any value of V_{GS}, but the curve is not always available. The endpoints for the curve can, however, be easily determined, as was shown in Chapter 4 and in the equation used.

If we choose a value of $V_{GS} = -2$ V, the drain current will be

$$I_D = 12\,\text{mA} \left(1 - \frac{-2}{-4}\right)^2 = 3\,\text{mA}$$

Of course, we should see this approximate value from the graph. What would the drain current be for V_{GS} values of -1.5 and -0.5 volts? _____

_ _ _ _ _ _ _ _ _ _ _ _ _ _ _ _ _ _

4.7 mA and 9.2 mA

Note: Manufacturers give I_{DSS} and $V_{GS(off)}$ over a wide range for a given JFET. Typical values are sometimes given and can serve as a starting point. You may have to resort to actually measuring them as mentioned in the frame.

33. Let us choose to set V_{GS} at -2 V. From frame 32, we see that the JFET will conduct a drain current of 3 mA. Now we must choose the value of drain to source voltage, V_{DS}. This will serve as the output voltage of our common source amplifier. If we choose a value of 10 V and use a supply of 24 V, then we can calculate the value of the required load resistance, in this case R_D.

$$R_D = \frac{(V_{DD} - V_{DS})}{I_D}$$

What would be the value of R_D necessary to set the output voltage at the level given in this frame? _____

_ _ _ _ _ _ _ _ _ _ _ _ _ _ _ _ _

$$R_D = \frac{14 \text{ V}}{3 \text{ mA}} = 4.67 \text{ k}\Omega$$

34. To use the JFET as an amplifier, we need to apply an input signal to the gate. This causes voltage changes on the gate, which result in corresponding drain current changes. For example, apply a sine wave of $0.5 \ V_{pp}$ to the gate as shown.

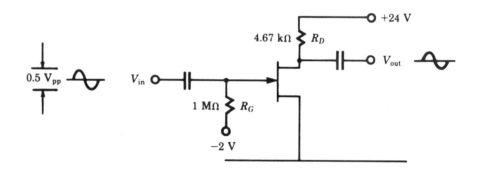

This means the gate voltage will vary from -1.75 to -2.25 V. What will the corresponding drain currents be? _____

_ _ _ _ _ _ _ _ _ _ _ _ _ _ _ _

For $V_{GS} = -1.75$ V, we can calculate $I_D = 3.8$ mA.
For $V_{GS} = -2.25$ V, we can calculate $I_D = 2.3$ mA.

35. The drain currents calculated for frame 34 will now cause a change in the voltage drop across resistor R_D. What will be the variation in this voltage? _____

_ _ _ _ _ _ _ _ _ _ _ _ _ _ _ _

For $I_D = 3.8$ mA, $V_{RD} = 17.7$ V.
For $I_D = 2.3$ mA, $V_{RD} = 10.7$ V.
This would be a voltage change of 7 V_{pp}.

36. The output voltage will also vary. What will be the variation in the output voltage? _____

_____

_____

_ _ _ _ _ _ _ _ _ _ _ _ _ _ _ _ _

For $I_D = 3.8$ mA, $V_{out} = 24 - 17.7 = 6.3$ V.
For $I_D = 2.3$ mA, $V_{out} = 24 - 10.7 = 13.3$ V.
This is also a voltage change of 7 V_{pp}.

37. The following table summarizes the calculations just made. The DC bias point is included.

V_{GS}	I_D	V_{RD}	V_{out}
−1.75 V	3.8 mA	17.7 V	6.3 V
−2.0 V	3.0 mA	14.0 V	10.0 V
−2.25 V	2.3 mA	10.7 V	13.3 V

The variation at the output represents an AC signal riding on top of a 10 V DC level. What are some characteristics of the AC output voltage?

_____

_____

_____

_ _ _ _ _ _ _ _ _ _ _ _ _ _ _ _ _

The output would have a value of 7 V_{pp} and would be a sine wave of the same frequency as the input sine wave. We should also note that as the input voltage on V_{GS} increases (toward 0 V), the output decreases. As the input voltage decreases (becomes more negative), the output voltage increases. This means that the output is 180° out of phase with the input. The amplifier has a 180° phase inversion from the input to the output.

38. The voltage gain for this amplifier could be found from

$$A_v = \frac{-V_{out}}{V_{in}}$$

The negative sign indicates 180° phase inversion. What would be the gain for the amplifier just discussed? _____

_ _ _ _ _ _ _ _ _ _ _ _ _ _ _

$A_v = -14$

39. Another way to calculate the gain is by using the expression

$$A_v = -(g_m)(R_D)$$

where g_m is called the transconductance and is a dynamic characteristic of the JFET. It is also called the forward transfer admittance. A typical value for g_m is generally provided for each JFET by manufacturers' specification sheets. It can also be found, however, from the data shown in the table of frame 37. The formula is

$$g_m = \frac{\Delta I_D}{\Delta V_{GS}}$$

where "Δ" indicates the change or variation in V_{GS} and the corresponding drain currents. The unit for transconductance is mhos or Siemens and is a very small number. The prefix for micro is usually attached to the value.

(a) Using the data from the table in frame 37, what is the value of g_m for the JFET used in the amplifier?

_____

(b) What is the corresponding voltage gain?

_____

– – – – – – – – – – – – – – – – –

(a) $g_m = \dfrac{1.5 \text{ mA}}{0.5 \text{ V}} = 0.003$ mhos

This would be expressed as 3000 μmhos.

(b) $A_v = -(0.003)(4670) = -14$, the same as before

40. Design a JFET common source amplifier using a JFET with the characteristics of $I_{DSS} = 14.8$ mA and $V_{GS(off)} = -3.2$ V. The input signal is 40 mV$_{pp}$. The drain supply is 24 V. Use the following steps:

(a) Establish the operating point at a value of voltage near the middle of the transfer curve.

_____

(b) Calculate the corresponding drain current using the formula in frame 32.

_____

(c) Assume a value of V_{DS} and calculate the value of R_D.

_____

(d) Let the gate supply vary according to the input signal and calculate the change in drain current.

_____

_____

(e)   Calculate the change in output voltage using the procedure of frames 35 and 36.

_____

_____

(f)   Calculate the gain of the amplifier.

_____

— — — — — — — — — — — — — — — —

(a)   assume a value of $V_{GS} = -1.6$ V
(b) $I_D = 3.7$ mA
(c)   assume $V_{DS} = 10$ V, $R_D = \dfrac{14V}{3.7\ mA} = 3780$ ohms
(d) V_{GS} will vary from -1.58 to -1.62 V; use the formula to calculate values of drain current. I_D will vary from 3.79 to 3.61 mA.
(e) V_{RD} will vary from 14.3 to 13.6 V and therefore V_{out} will vary from 9.7 to 10.4 V.
(f) $A_v = \dfrac{-0.7}{0.04} = -17.5$

41.   From the results of step (d), calculate the transconductance of the JFET and the gain from $A_v = -g_m R_D$.

_____

— — — — — — — — — — — — — — — —

$g_m = \dfrac{0.18\ mA}{40\ mV} = 4500\ \mu mhos$

$A_v = -(0.0045)(3780) = -17$, which is very close to the value found in frame 40, step (f).

42.   It is also possible to use one power supply and a self biasing technique to operate the JFET. Such a circuit is shown here.

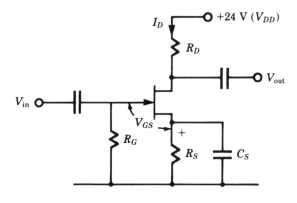

Since the gate doesn't have any current flow, the voltage drop across R_S becomes the gate to source voltage. To design the circuit we will need to find values for both R_S and R_D. Again select the bias point of $V_{GS} = -2$ V and $I_D = 3$ mA as in frames 33 through 37. The value of R_S can now be found, recognizing that $V_S = V_{GS}$ in magnitude.

$$R_S = \frac{V_{RS}}{I_D} = \frac{V_{GS}}{I_D}$$

The value calculated for R_S will set up $V_{GS} = -2$ V. Now if V_{DS} is maintained at 10 V, then resistor R_D can be calculated.

$$R_D = \frac{(V_{DD} - V_{DS} - V_S)}{I_D}$$

The value of C_S is chosen so that its reactance is less than 10% of R_S at the lowest frequency to be amplified. The DC load for the JFET will be R_D plus R_S. The AC load will be R_D only, since C_S is bypassing the AC signal around R_S. This keeps the DC operating point stable. It reduces the gain slightly since a smaller R_D is now used to calculate the AC voltage swings at the output.

(a)   What value should R_S be? _____

(b)   What value should R_D be? _____

(c)   What value should C_S be? Assume $f = 1$ kHz. _____

_____

(d)   Calculate V_{out} for the same V_{in} of frame 34, 0.5 V_{pp}.

_____

_____

_____

(e)   What is the voltage gain? _____

- - - - - - - - - - - - - - -

(a) $R_S = \dfrac{2\text{ V}}{3\text{ mA}} = 667$ ohms

(b) $R_D = \dfrac{12\text{ V}}{3\text{ mA}} = 4$ kΩ

(c)   make $X_{CS} = 66.7$ ohms, $C_S = 2.4$ μF

(d)   The AC drain current will still vary from 3.8 to 2.3 mA (see frame 37). The voltage across R_D will now be 6 V_{pp} since R_D is now 4 kΩ. The output voltage will also be 6 V_{pp}.

(e) $A_v = \dfrac{-6}{0.5} = -12$; the gain will be 12

THE OPERATIONAL AMPLIFIER

43. The operational amplifier (op-amp) in use today is actually an integrated circuit (IC). This means that the device has numerous solid state components and resistors all constructed on the same silicon substrate and packaged in a form that is very small and easily utilized. It is possible to purchase op-amps in different case configurations. Some of these configurations are: the TO metal package; the flat pack; the 14 pin DIP; and the mini DIP. It is also possible to have op-amps packaged with 2 op-amps (dual) or 4 op-amps (quad) in the same IC package. Their size, low cost, and wide range of application have made op-amps so common today that they are thought of as a circuit device or component in themselves, even though a typical op-amp may contain 20 or more transistors in its design. The characteristics of op-amps very closely approach those of an ideal amplifier. These characteristics are:

- high input impedance (does not require input current);

- high gain (used for amplifying small signal levels);

- low output impedance (not affected by the load).

(a) What are the advantages of using op-amps? _____

_____

_____

(b) Why are op-amps packaged as ICs? _____

_____

_____

- - - - - - - - - - - - - - - - - -

(a) small size, low cost, wide range of applications, high input impedance, high gain, and low output impedance
(b) The large numbers of transistors and components that are required in the design of an op-amp make them suitable to be packaged in a small IC package. This saves a lot of space.

44. The general symbol for the op-amp looks like this:

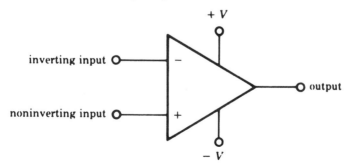

An input at the inverting input will cause an output that is 180° out of phase with the input. An input at the noninverting input will cause an output that is in phase with the input. Both positive and negative supplies will be required, and their values will be specified for the particular op-amp used. Other terminals will sometimes be shown for different applications. It is the task of the designer to arrange the external components and connections to make the op-amp and associated circuitry perform the function required. Numerous handbooks and applications manuals that are available to describe op-amp circuits are very useful to the designer or the hobbyist.

(a) How many terminals does the op-amp require, and what are their functions? _____

_____

(b) How is the output related to the input when the input is connected to the inverting input? _____

_____

– – – – – – – – – – – – – – – –

(a) 5—two input terminals, one output terminal, two power supply terminals
(b) The output is 180° out of phase with the input.

45. The most basic op-amp circuit is shown here. It utilizes a type 741 op-amp as an inverting amplifier.

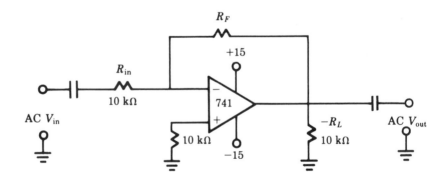

The voltage gain for the circuit is found from the equation

$$A_v = \frac{-R_F}{R_{in}}$$

Resistor R_F is called a feedback resistor, because it forms a feedback path from the output to the input. Many op-amp circuits involve the use of a feedback loop. Since the op-amp has such a high gain, it is easy to saturate it (at maximum gain) with only very slight changes in potential between the two input terminals. The feedback loop allows the operation of the op-amp at lower gains over a wider range of input differences. The designer can choose the value of the feedback resistor to achieve a certain voltage gain. The capacitors in the diagram are to keep DC voltages out of the input and output.

(a)   If the desired gain for the circuit is 120, what should the value of

R_F be in the circuit shown? _____

(b)   If the AC V_{in} is 5 V_{rms}, what will the output voltage be?

_____

— — — — — — — — — — — — — — — —

(a) $R_F = 120 \times 10\ k\Omega = 1.2\ M\Omega$
(b) $V_{out} = 120 \times 5\ mV = 0.6\ V_{rms}$; the output signal will also be inverted with respect to the input signal

46.   For the op-amp circuit shown in frame 45, the values of the input resistors are 6.8 kΩ. The desired output is a voltage of 12 V_{pp}. The designer decides to utilize a gain of 50, which he or she knows from the manufacturer's specification sheets will produce a proper op-amp frequency response at the frequencies the designer plans to use.

(a)   What value of R_F is required? _____

(b)   What should the input be set at to get the required output?

_____

— — — — — — — — — — — — — — — —

(a) $R_F = 50 \times 6.8\ k\Omega = 340\ k\Omega$
(b) $V_{in} = \dfrac{12\ V_{pp}}{50} = 0.24\ V_{pp} = 0.168\ V_{rms}$

SUMMARY

This chapter has introduced the most common types of amplifiers in use today: the common emitter BJT, the common source JFET, and the op-amp. The material presented, at best, only scratches the surface of the world of amplifiers. There are many variations and types of amplifiers as well as other types of devices that will perform amplification. The terminology and design approach used here should form a basic foundation for further learning.

SELF-TEST

The most important things you learned in this chapter were:

(1) How to design a simple amplifier when the bias point and the gain were specified.
(2) How to do the same for an emitter follower.
(3) How to analyze a simple circuit.

The questions below will test your understanding of this chapter. Use a separate sheet of paper for your diagrams or calculations. Compare your answers with the answers provided following the test.

1. What is the main problem with the original amplifier circuit shown in frame 1? _____

_____

2. What is the gain formula for that circuit? _____

3. Does it have a high or low gain? _____

Use the circuit shown here for questions 4–8.

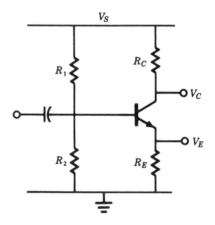

4. Design an amplifier so that the bias point is 5 V and the AC gain is 15. Assume $\beta = 75$, $R_{in} = 1.5$ kΩ, $V_S = 10$ V, and $R_C = 2.4$ kΩ. Add capacitor C_E and calculate a suitable value to maintain maximum gain at 50 Hz. What is the approximate value of this gain? _____

_____

5. Repeat question 4 with these values: $V_S = 28$ V, $\beta = 80$, $R_{in} = 1$ kΩ, and $R_C = 10$ kΩ. The bias point is to be 14 V, and the gain 20.

_____

_____

6. Repeat question 4 with these values: $V_S = 14$ V, $\beta = 250$, $R_{in} = 1$ kΩ, and $R_C = 15$ kΩ. The bias point is to be 7 V, and the gain 50.

_____

_____

7. Design an emitter follower.
   $V_S = 12$ V, $R_E = 1$ kΩ, $\beta = 100$, $V_E = 6$ V, $R_C = 0$ ohms

   _____

8. Design an emitter follower.
   $V_S = 28$ V, $R_E = 100$ ohms, $\beta = 35$, $V_E = 7$ V, $R_C = 0$ ohms

   _____

In questions 9–11, the resistance and β values are given. Analyze the circuit. That is, find the bias point and the gain.

9. $R_1 = 16$ kΩ, $R_2 = 2.2$ kΩ, $R_E = 100$ ohms, $R_C = 1$ kΩ, $\beta = 100$,
   $V_S = 10$ V _____

10. $R_1 = 36$ kΩ, $R_2 = 3.3$ kΩ, $R_E = 110$ ohms, $R_C = 2.2$ kΩ, $\beta = 50$,
    $V_S = 12$ V _____

11. $R_1 = 2.2$ kΩ, $R_2 = 90$ ohms, $R_E = 20$ ohms, $R_C = 300$ ohms, $\beta = 30$,
    $V_S = 50$ V _____

12. The circuits from questions 4 and 5 are cascaded. What is the gain when both have a capacitor around the emitter? When the capacitor is not

    used in either of them? _____

    _____

13. Design a JFET amplifier using the circuit of frame 34. The characteristics of the JFET are $I_{DSS} = 20$ mA and $V_{GS(off)} = -4.2$ V. The desired value of V_{DS} is 14 V. Find the value of R_D.

    _____

14. If the transconductance of the JFET used in problem 13 is 4800 μmhos, what will the voltage gain be? _____

    _____

15. If the desired output is 8 V_{pp} for the JFET of problems 13 and 14, what should the input be? _____

16. Design a JFET amplifier using the circuit of frame 42. The JFET characteristics are $I_{DSS} = 16$ mA and $V_{GS(off)} = -2.8$ V. Using a V_{DS} of 10 V, find the values of R_S, C_S, and R_D.

_____

_____

_____

17. If the input to the JFET of problem 16 is 20 mV$_{pp}$, what will the AC

   output voltage be and what is the gain? _____

_____

_____

_____

18. In the op-amp of frame 45, what will the output voltage be if the input

   is 50 mV and the feedback resistor is 750 kΩ? _____

_____

Answers to Self-Test

If your answers do not agree with those given below, review the frames indicated in parentheses before you go on to the next chapter.

1. Its bias point is unstable, and its gain varies with temperature. Also you cannot guarantee what the gain will be. (text on page 180)

2. $A_V = \beta \times \dfrac{R_C}{R_{in}}$ (frame 10)

3. Usually the gain is quite high. (text on page 180)

For numbers 4–6, suitable values are given. Yours should be close to these.

4. $R_1 = 29$ kΩ, $R_2 = 3.82$ kΩ, $R_E = 160$ ohms, $C_E = 200$ μF, $A_V = 120$
   (frames 17–22)

5. $R_1 = 138$ kΩ, $R_2 = 8$ kΩ, $R_E = 500$ ohms, $C_E = 64$ μF, $A_V = 800$
   (frames 17–22)

6. $R_1 = 640$ kΩ, $R_2 = 45$ kΩ, $R_E = 300$ ohms, $C_E = 107$ μF, $A_V = 750$
   (frames 17–22)

7. $R_1 = 8$ kΩ; $R_2 = 11.2$ kΩ (frame 27)

8. $R_1 = 922$ ohms; $R_2 = 385$ ohms (frame 27)

9. $V_C = 5$ V, $A_V = 10$ (frames 28–30)

10. $V_C = 6$ V, $A_V = 20$ (frames 28–30)

11. $V_C = 30$ V, $A_V = 15$ (frames 28–30)

12. When the capacitor is used, $A_V = 120 \times 800 = 96{,}000$.
    When the capacitor is not used, $A_V = 15 \times 20 = 300$.

              (frames 17–22)

13. Assume $V_{GS} = -2.1$ V, $I_D = 5$ mA, $R_D = 2$ kΩ   (frames 31–33)

14. $A_v = -9.6$ (frame 39)

15. $V_{in} = 83$ mV$_{pp}$ (frame 38)

16. Use $V_{GS} = -1.4$ V, then $I_D = 4$ mA
    $R_S = 350$ ohms
    $C_S = 4.5$ μF (assume $f = 1$ kHz)
    $R_D = 3.15$ kΩ   (frame 42)

17. V_{GS} will vary from -1.39 to -1.41 V, I_D will vary from 4.06 to 3.94 mA,
    V_{out} will be 400 mV$_{pp}$, $A_v = \dfrac{-400}{20} = -20$ (frame 42)

18. $A_v = -75$, $V_{out} = 3.75$ V   (frame 45)

CHAPTER NINE

Oscillators

The purpose of this chapter is to introduce and discuss the basis of oscillators. It will not provide a complete coverage of the subject.

An oscillator is a circuit which produces a continuous output signal; thus it is called a *signal generator*. When the signal produced is a sine wave of constant amplitude and frequency, the oscillator circuit is called a *sine wave generator*. Radio and TV signals are sine waves transmitted through the air, the 120 V AC from the wall plug is a sine wave, as are many test signals used in electronics.

In this chapter we introduce three basic sine wave oscillators. They all rely on the *LC* tuned circuit described in Chapter 7. There are many types of oscillator circuits which are used extensively in practical electronics. Other types of oscillators produce a variety of different output signals, such as square waves, triangle waves, and pulses. Since most of these are the domain of digital electronics they are not covered here.

When you complete this chapter, you will be able to:

- recognize the main elements of an oscillator;

- differentiate between positive and negative feedback;

- specify the type of feedback needed for an oscillator;

- specify at least two methods of obtaining feedback in an oscillator circuit;

- identify the feature which governs the frequency of an oscillator;

- design a simple oscillator circuit.

1.   An oscillator can be divided into three definite sections: (1) an amplifier; (2) the feedback connections; and (3) the frequency determining components.

An *amplifier* can be converted into an oscillator by using a resonant *LC* circuit as the *frequency determining* components and by incorporating *feedback* from the output to the input.

The amplifier replaces the switch in the circuit shown in frame 35 of Chapter 7. It is more convenient to consider it as an amplifier than a switch since amplification is necessary in an oscillator.

Draw the circuit, and label the parts.

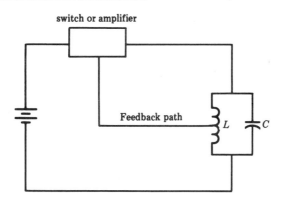

2. Feedback is provided when we connect the output of an amplifier to its input. If the output fed back is "out of phase" with the input, then the circuit has *negative feedback* (NFB).

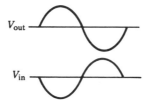

An example of negative feedback would be connecting the collector of a transistor amplifier to its base, through a feedback resistor, R_f, as shown below.

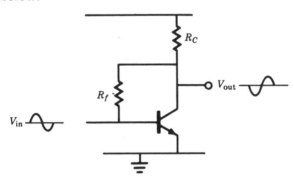

The effect of negative feedback is to stabilize the operation of an amplifier. This means it does the following.

Prevents the DC bias point drifting with temperature and component changes.

Prevents gain changes caused by the same reasons. It also reduces the gain, but makes it easier to control the gain.

Reduces distortion in amplifiers, thus improving the "quality" of the sound.

(a) Do high quality hi-fi amplifiers have feedback? _____

(b) If so, what kind of feedback do they have? _____

_ _ _ _ _ _ _ _ _ _ _ _ _ _ _ _ _ _

(a) yes, always
(b) negative feedback

3. If the feedback from the output is in phase with the input, the circuit has *positive feedback*.

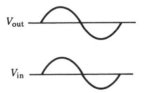

An example would be connecting the collector of the second transistor in a two-transistor amplifier to the base of the first transistor. This causes oscillations.

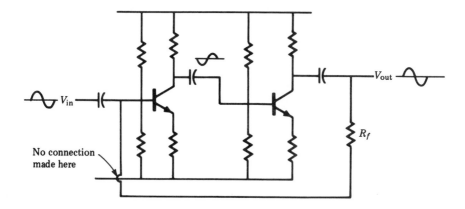

Positive feedback will cause an amplifier to oscillate even when there is *no external input.*

(a)   What type of feedback is used to stabilize an amplifier? _____

(b)   What type of feedback is used in oscillators? _____

(c)   What general type of connection produces feedback? _____

_____

─ ─ ─ ─ ─ ─ ─ ─ ─ ─ ─ ─ ─ ─ ─ ─ ─

(a)   negative feedback
(b)   positive feedback (PFB)
(c)   connecting the output of an amplifier to its input

4.   The amplifier shown below is the same as those we discussed in Chapter 8. It is called a common emitter amplifier circuit.

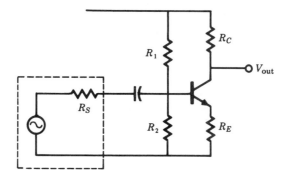

It has a simple resistive load, R_C.

(a)   What would be the effect of negative feedback on this amplifier?

_____

(b) What would be the effect of positive feedback? _____

_____

- - - - - - - - - - - - - - - - -

(a) stabilize it, reduce gain, and reduce distortion
(b) cause it to oscillate

5. An oscillator circuit must have a large gain, which is why an amplifier is needed. In the circuit in frame 4, if an external signal is applied to the base it will be amplified.

(a) What is the basic formula for an amplifier's gain?

_____

(b) What is the gain formula for the above amplifier?

_____

- - - - - - - - - - - - - - - - -

(a) $A_V = \beta \times \dfrac{R_L}{R_{in}}$

(b) $A_V = \dfrac{R_L}{R_E} = \dfrac{R_C}{R_E}$

(since R_C is the only load in this circuit)

6. It is also possible to provide an input to the emitter of an amplifier instead of the base. This is called a *common base* circuit.

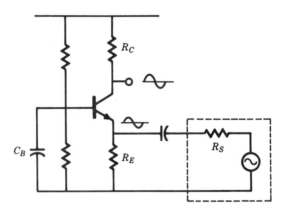

The gain formula is still the same basic amplifier gain formula, but here the input impedance to the amplifier is very low when the signal is fed into the emitter. In practice, the basic gain formula simplifies to the following.

$$A_V = \frac{R_L}{R_S}$$

R_S is the output resistance or impedance of the source or generator. It may also be called the *internal impedance* of the source. (This formula can be verified mathemat.cally, but we will not do this here.)

What would be the actual gain formula for the above circuit?

_ _ _ _ _ _ _ _ _ _ _ _ _ _ _ _

$$A_V = \frac{R_C}{R_S}, \text{ (since } R_C \text{ is the load)}$$

7. Notice that the input and output sine waves in the figure in frame 6 are all in phase. Although the signal is amplified, it is not inverted. This has practical applications when building oscillators.

(a) What happens to the input signal to the amplifier when it is applied to the emitter instead of the base? _____

(b) Is the input impedance of the common base amplifier high or low compared to the common emitter amplifier? _____

(c) What is the gain formula for the common base amplifier?

_____

_ _ _ _ _ _ _ _ _ _ _ _ _ _ _ _

(a) amplified and not inverted
(b) low

(c) $A_V = \dfrac{R_L}{R_S}$

8. Suppose now we modify this amplifier and use a parallel LC circuit connected as shown below. This is sometimes called a *tuned*, or *resonant*, *load*.

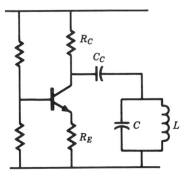

Component C_C is included to avoid a DC connection between the collector and the inductor. The inductor has a very small DC resistance, which would pull the collector DC voltage down to near 0 V and stop the circuit from behaving as an amplifier. The size or value of C_C in oscillators will be covered shortly.

(a) What term would you use to describe the load here? _____

(b) Does the circuit have all three components of an oscillator yet?

_____

─ ─ ─ ─ ─ ─ ─ ─ ─ ─ ─ ─ ─ ─

(a) resonant or tuned
(b) no, the feedback connections are missing

9.   The gain of the amplifier with the resonant load is slightly different. Its total load is now the parallel combination of the collector resistor R_C and the tuned LC circuit. The exact value of this depends on the input frequency, and is rather difficult to calculate.

At the resonant frequency of the tuned load, the LC circuit has a very high impedance. The value of R_C is chosen to be less than the impedance of the tuned circuit at resonance, so that at the resonant frequency the load becomes approximately R_C.

Refer back to the circuits and basic gain formulas, if necessary, and write the gain formula for the following.

(a) common emitter circuit _____

(b) common base circuit _____

─ ─ ─ ─ ─ ─ ─ ─ ─ ─ ─ ─ ─

(a) $A_V = \dfrac{R_C}{R_E}$

(b) $A_V = \dfrac{R_C}{R_S}$

10. Both the common emitter and the common base circuits are used in oscillators, and in each case an extra capacitor is usually included.

In the common emitter amplifier, an emitter capacitor C_E is added. This is similar to the emitter capacitor discussed in the chapter on amplifiers.

In the common base circuit a capacitor C_B is used to connect the base to ground. This was shown in frame 6.

What is the general effect in both cases? _____

_____

─ ─ ─ ─ ─ ─ ─ ─ ─ ─ ─ ─ ─

an increase in the gain of the amplifier

The gain is increased to the point where it can simply be considered "large enough" to enable the amplifier to be used as an oscillator. When these capacitors are used, it is not usually necessary to calculate the gain of the amplifier.

11. The frequency at which an oscillator will oscillate is the resonant frequency of its LC circuit. What is the formula for the oscillation (or

resonant) frequency? _____

— — — — — — — — — — — — — — — — —

$$f_r = \frac{1}{2\pi\sqrt{LC}}$$

In practice the actual measured frequency is never quite the same as the calculated frequency. The capacitor and inductor values are not exact, and other stray capacitances in the circuit will affect the frequency. When accuracy is required in operation, an adjustable capacitor or inductor is used.

12. An alternative method of connecting the tuned load is shown below.

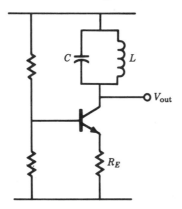

This and the circuit in frame 8, both tuned amplifiers, are used extensively in radio and TV circuits, since they will amplify one frequency far more than all other frequencies.

(a) What would you expect this one frequency to be? _____

(b) Write the formula for the impedance of the circuit at the resonant

frequency. _____

(c) What is the gain at this frequency? _____

— — — — — — — — — — — — — — — —

(a)   the resonant frequency

(b) $Z = \dfrac{L}{C \times r}$. Where r is the DC resistance of the coil (In practice the value of Z becomes large.)

(c) $A_V = \dfrac{Z}{R_E}$. Or, just "large."

13.   Due to the very low resistance of the coil, the DC voltage at the collector is usually very close to the supply voltage V_S. In addition, the AC output voltage positive peaks can exceed the DC level of the supply voltage. With large AC output, the positive peaks can actually reach $2V_S$.

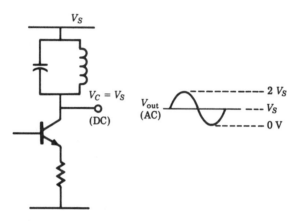

Indicate which of the following is an accurate description of the circuit above.

_____ (a)   oscillator   _____ (c)   common base circuit

_____ (b)   tuned amplifier   _____ (d)   common emitter circuit

- - - - - - - - - - - - - - - - - - -

b

FEEDBACK

14.   To convert an amplifier into an oscillator, a portion of the output signal must be connected to, or fed back, to the input. The feedback signal must have the correct phase to induce oscillations.

The feedback is usually taken from the tuned load. Three main methods are used; each is a variation on the voltage divider circuit. These are shown in the following diagrams. (Each is named for the original inventor.)

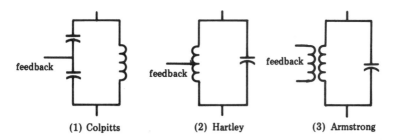

(1) Colpitts   (2) Hartley   (3) Armstrong

The Colpitts method uses a capacitive voltage divider. Remember, both Colpitts and Capacitor begin with a C. The Hartley method uses a "tap" on the coil to provide an inductive voltage divider. The Armstrong method uses an extra winding, usually with fewer turns than the main winding. This is in fact a transformer rather than a true voltage divider. In all three types, between one tenth and a half of the output must be used as feedback.

(a) Where is the feedback taken from in a Colpitts oscillator?

_____

(b) What type of oscillator uses a tap on the coil for the feedback

voltage? _____

(c) What type does not use a voltage divider? _____

– – – – – – – – – – – – – – – –

(a) a capacitive voltage divider
(b) Hartley
(c) Armstrong

15. The output voltage appears at one end of the circuit shown below, and the other end is effectively at ground. The feedback voltage is taken between the junction of the two capacitors and ground; it is labeled V_f.

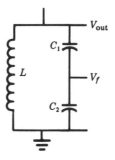

Using the voltage divider formula, what is V_f? _____

– – – – – – – – – – – – – – –

$$V_f = \frac{V_{out} X_{C2}}{(X_{C_1} + X_{C_2})}$$

which becomes $V_f = \dfrac{V_{out} C_1}{(C_1 + C_2)}$

16. In a resonant circuit of this form, how is the resonant frequency found? First, the equivalent total capacitance C_T of the two series capacitors must be found. C_T is then used in the resonant frequency formula.

(a)   What is the formula for C_T? _____

(b)   What is the resonant frequency formula for the Colpitts oscillator?

_____

– – – – – – – – – – – – – – – –

(a) $C_T = \dfrac{C_1 C_2}{C_1 + C_2}$

(b) $f_r = \dfrac{1}{2\pi\sqrt{LC_T}}$, if Q is less than 10

Note: The other formulas given in Chapter 7 for parallel resonance should also be looked at in making the calculation, since Q may be small. These formulas are repeated here.

$$f_r = \frac{1}{2\pi\sqrt{LC}} \sqrt{1 - \frac{r^2 C}{L}} = \frac{1}{2\pi\sqrt{LC}} \sqrt{\frac{Q^2}{1 + Q^2}}$$

17. The feedback voltage is taken from a tap on the coil, as shown in the circuit below.

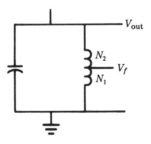

In the diagram, N_1 and N_2 are the number of turns in each of two sections of the coil. The voltage out can be found from the voltage divider formula modified to include the number of turns in each part of the coil.

$$V_f = V_{out} \times \frac{N_1}{N_1 + N_2}$$

The practical difficulty here is that often the number of turns is not known, and neither is the "turns ratio," since the exact place of the tap may not be known. This problem can be especially acute in audio work and low frequency circuits where a large number of turns is used. Usually when this method is used, the number of turns in the coil is specified by the manufacturer.

(a) Who is the inventor of this method? _____

(b) What amount of V_{out} should V_f be? _____

- - - - - - - - - - - - - - - -

(a) Hartley
(b) Between 1/10 and 1/2

18. As the feedback is taken from the secondary winding in a transformer the output voltage is easily calculated. This will be covered in Chapter 10, frame 6, so we will not pursue it further here.

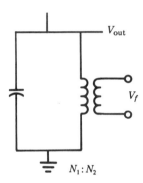

Who invented this type of oscillator? _____

- - - - - - - - - - - - -

Armstrong

19. In all of the above cases the voltage fed back from the output by the voltage divider is a fraction of the total output voltage. This V_f fraction is always less than 1.

In order to ensure oscillations the product of the feedback and the amplifier gain must be greater than 1.

$$A_V \times V_f > 1$$

It is usually easy to achieve this since A_V is much greater than 1.,
No external input is applied to the oscillator. Its input is the small part of the output signal which is fed back. If this feedback is of the correct phase and amplitude, the oscillations will start spontaneously and will continue as long as power is supplied to the circuit.

The transistor amplifier amplifies the feedback to sustain the oscillations, and it converts the DC power from the battery or power supply into the AC power of the oscillations.

(a)   What makes an amplifier into an oscillator? _____

_____

(b)   What input does an amplifier need to become an oscillator?

_____

_____

– – – – – – – – – – – – – – – –

(a)   a tuned circuit with feedback of the correct phase and amount
(b)   none—oscillations will happen spontaneously if the feedback is correct

If you wish to take a break, this is a good stopping point.

THE COLPITTS OSCILLATOR

20.   The Colpitts is the simplest of the LC oscillators to build.

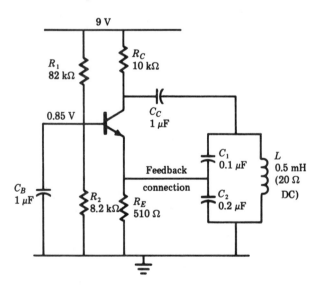

The feedback is taken from the capacitive voltage divider and is fed to the emitter. This connection provides the correct phase for the feedback.

The capacitor C_B is chosen such that its reactance will be low enough that the AC signal sees the capacitor as a path, rather than R_2, at the lowest frequency that is to be sustained in the oscillator. For a low-range oscillator, choose a frequency of 1 kHz and a reactance of 160 ohms and then check the reactance at the frequency of oscillation to be sure it is proper. If R_2 happens to be smaller than 1.6 kΩ, then choose a value of X_{CB} that is less than one tenth of R_2. The actual value of C_B has only to be large enough to offer a low reactance bypass that will be sufficient to keep the DC bias point constant.

For the circuit shown in this frame, what would be your first estimate for C_B? _____

- - - - - - - - - - - - - - - - - -

$$X_C = 160 \text{ ohms} = \frac{1}{2 \times \pi \times 10^3 \times C_B}$$

So $C_B = 1\ \mu F$. Thus, C_B can be $1\ \mu F$ or larger.

21. The component values used in frame 20 will produce a working oscillator. Work through the following questions as you analyze the circuit.

(a) What is the effective total capacitance of the two series capacitors in the tuned circuit? $C_T =$ _____

(b) What is the oscillator frequency? $f_r =$ _____

(c) What is the approximate impedance of the tuned circuit at this frequency? $Z =$ _____

(d) What fraction of the output voltage is fed back? $V_f =$ _____

(e) What is the reactance of C_B at the frequency of oscillation?

$X_{CB} =$ _____

- - - - - - - - - - - - - - - - - -

(a) 0.067 μF

(b) Since Q is not known, use the formula that includes the resistance of the coil. (See frame 16.) $f_r = 26.75$ kHz. This actually means that the coil has a low Q of 4.2.

(c) Using $Z_T = \frac{L}{rC}$, $Z_T = 373$ ohms.

(d) Using a voltage divider with the capacitor values,

$$V_f = V_{out} \frac{C_1}{(C_1 + C_2)} = \frac{V_{out}}{3}.$$

(e) $X_{CB} =$ about 6 ohms, which is a very good value, much less than the 8200 ohm value of R_2.

22. In the circuit in frame 20 note how easy it is to adjust the frequency and the feedback voltage by changing the capacitors.

    In practice these would be adjusted to find the largest and the smallest ratio at which oscillations would occur and then a value about half way between would be used. A variation on this circuit is shown below.

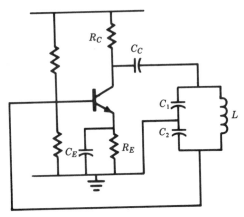

What is different in this circuit? _____

_____

- - - - - - - - - - - - - - - -

The feedback is connected to the base instead of the emitter, and the ground connection is different. The capacitor C_E is added. (This connection provides the correct phase of feedback to the base.)

23. Capacitor C_E is chosen so that it has a reactance of less than 160 ohms at the desired frequency. If the emitter resistor R_E is smaller in value than 1 kΩ, then C_E should be re-chosen so that its reactance is less than $R_E/10$ at the oscillation frequency.

    If an emitter resistor of 510 ohms is used in a 1 kHz oscillator,

what value of capacitor should be used for C_E? _____

- - - - - - - - - - - - - - - -

$$X_c = \frac{510}{10} = \frac{1}{2\pi f C_E} = \frac{0.16}{10^3 \times C_E}$$

So $C_E = 3.2 \ \mu F$. Thus, a capacitor larger than 3 μF should be used.

24. An alternative connection for the tuned load is shown in the circuit on the next page.

The inductor used in this case is a small commercially available audio transformer, typical of the sort which might be used in a lab or workshop to make a simple audio oscillator. (It is a Calectro D1-722.) The small letters G and R indicate the colors of the leads (green and red), which you need to know when hooking it up.

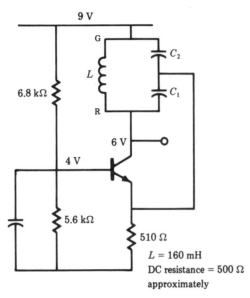

The different values for the capacitors are shown below.

C_1	C_2	C_T	f_r
0.01 μF	0.1 μF	_____	_____
0.01 μF	0.2 μF	_____	_____
0.01 μF	0.3 μF	_____	_____
0.1 μF	1 μF	_____	_____
0.2 μF	1 μF	_____	_____

(a) Calculate and fill in the value of C_T and f_r for each case.

(b) Does increasing C_2 while holding C_1 constant act to increase or decrease the frequency? _____

(c) What effect does increasing C_1 have? _____

(d) What is the condition that results in the highest possible frequency? _____

(e) What would be the highest frequency if C_1 is fixed at 0.01 μF and C_2 can vary from 0.005 μF to 0.5 μF? _____

(a)   You should have approximately these values.

C_T	f_r
0.009 μF	4.19 kHz
0.0095 μF	4.08 kHz
0.0097 μF	4.04 kHz
0.09 μF	1.33 kHz
0.167 μF	0.97 kHz

(b)   Increasing C_2 will decrease the frequency of oscillation.
(c)   Increasing C_1 will also decrease the frequency.
(d)   When C_T is at its lowest possible value.
(e)   When C_2 is 0.005 μF, C_T will be 0.0033 μF, which will be its lowest value; the frequency will be about 6.9 kHz. The lowest frequency would be when C_2 is at its highest setting of 0.5 μF.

Optional Experiment

Now build this oscillator in an experimental setup and see how close your measurements of the frequency agree with those you have just calculated. If you are within 20%, then the answers are satisfactory. In some cases the waveform may be distorted.

THE HARTLEY OSCILLATOR

25.   The feedback in the Hartley oscillator is taken from a tap on the coil. Here the wire lead colors are indicated as G (green), B (black), and W (white).

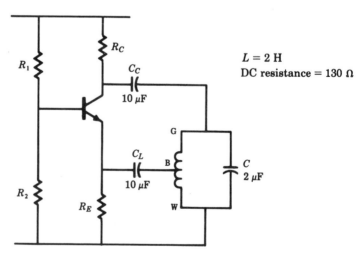

$L = 2$ H
DC resistance = 130 Ω

Capacitor C_L is necessary so that the low DC resistance of the coil will not pull the emitter DC voltage down to 0 V. C_L should have a reactance of less than $R_E/10$, or less than 160 ohms at the oscillator frequency.

Work through the following calculations.

(a)   What is the frequency? $f_r =$ _____

(b)   What is the approximate impedance
      of the load? $Z =$ _____

(c)   Why can't you calculate the fraction of the voltage that is fed

      back? _____

- - - - - - - - - - - - - - - - - -

(a)   80 Hz (approximately)
(b)   7.7 kΩ (approximately)
(c)   The number of turns and the turns ratio of the coil is not known.

Optional Experiment

An alternative Hartley circuit is shown below.

The coil is the same Calectro transformer used in frame 24, with a different winding used. Set up this circuit. The frequency will be about 1 kHz.

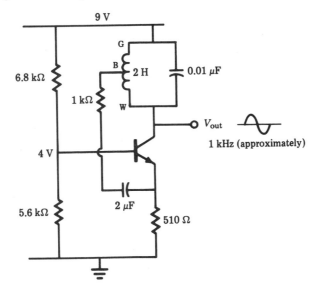

The two experiments give you a practical lesson. In the Colpitts oscillator it is easy to choose the two capacitors to provide both the desired frequency and the desired amount of feedback. If the circuit will not oscillate, it is relatively simple to change the capacitors until it does oscillate, and then adjust the values slightly to get the desired frequency.

In the Hartley it is not quite so easy. The feedback ratio cannot be altered, as it is impossible to make another tap or change the tap on the coil.

THE ARMSTRONG OSCILLATOR

The Armstrong oscillator is somewhat more difficult to design and build. A simple circuit is shown below.

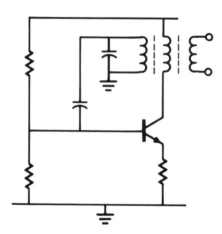

Here the oscillations depend more on the extra winding on the coil than on any other factor.

Because of the large variety of transformers and coils available it is almost impossible to give a definite simple procedure for designing an Armstrong oscillator. The manufacturer will specify the number of turns required on the coils, thus guaranteeing that the oscillator will work in its most common operation, at high radio frequencies.

Because of the practical difficulties, we will not pursue the Armstrong and its variations any further.

This is a good stopping point if you are ready for a study break.

PRACTICAL OSCILLATOR DESIGN

26. In this section we will briefly cover some practical problems with oscillators, and then present a simple design procedure.

    Before we do this, review the important points of this chapter by answering these questions.

    (a) An oscillator must have three factors present to work. List them.

    _____

    _____

    (b) How is the frequency determined? _____

(c) What provides the feedback? _____

(d) How many feedback methods for oscillators have we seen?

_____

(e) What is needed to start the oscillations once the circuit has been

built? _____

_ _ _ _ _ _ _ _ _ _ _ _ _ _ _

(a) an amplifier, a tuned load, and feedback
(b) by the resonant frequency of the tuned load
(c) a voltage divider on the tuned load
(d) three—the Colpitts, Hartley, and the Armstrong
(e) nothing—the oscillations should start spontaneously if the circuit
is correct

The main practical problem in building oscillators is selecting the coil. For mass production a manufacturer can specify and purchase the exact coil required. But in a lab or workshop, where only a single circuit is to be built, it is often difficult or impossible to find the exact inductor specified in a circuit design. So what usually happens is that the most readily available coil is used and the rest of the circuit is designed around it. This produces four main problems.

The exact value of the inductance may not be known.

The inductance value may not be the best for the frequency range desired.

The inductance value may change by a large amount over a narrow frequency range around the desired frequency. This is due to capacitance in the coils, and the current in the coil windings, etc.

The coil may or may not have tap points or extra windings, and this may cause a change in the original circuit desired. For example, if there are no taps then a Hartley oscillator cannot be built, and if there are no extra windings then an Armstrong cannot be built.

Because it is the easiest to make work in practice and provides an easy way around some of the practical difficulties, let's return to the Colpitts oscillator.

Almost any coil can be used when building a Colpitts oscillator, provided it is suitable for the frequency range desired. For example, a coil from the tuner section of a TV set would not be really suitable for a 1 kHz audio oscillator. Its inductance value is outside the range best suited to a low frequency audio circuit.

Simple Oscillator Design Procedure

27. The following is a simple step-by-step procedure for the design of a Colpitts oscillator. The Colpitts was chosen as it is the easiest to build and to make work, and it will work over a wide frequency range. (A Hartley can be built from a very similar set of steps.)

By following this procedure an oscillator can be built which will work in the majority of cases. A procedure which will guarantee that the oscillator will work is far more complex.

If you are not actually building the oscillator, use the assumed values given here, and draw the circuits using your calculated values.

The steps are as follows.

(1) Decide upon the oscillator frequency.
(2) Choose a suitable coil. This step presents the greatest practical difficulty. Often a desired value of coil will not be available, and one must use whatever is readily available. Fortunately a wide range of values can be used.
(3) If the value of the inductance is known, calculate the capacitor value using the formula below.

$$f_r = \frac{1}{2\pi\sqrt{LC}}$$

Use this value of capacitor for C_1 in the next steps.
(4) If the inductance value is not known, then choose any value of capacitance and call this C_1. This may produce a frequency considerably different from that required. But at this stage the main thing is to get the circuit oscillating. Values can be adjusted later.
(5) Choose capacitor C_2 to be between 3 and 10 times the value of C_1. The two capacitors and coil are connected as shown here.

After step 5, let's stop and give some assumptions. Suppose we need a frequency of 10 kHz and have a coil with a 16 mH inductance.

(a) What approximate value of C_1 will be needed? _____

(b) What value of C_2 is needed? _____

- - - - - - - - - - - - - - - - -

(a) $C_1 = 0.016 \ \mu F$
(b) $C_2 = 0.048 \ \mu F$ to $0.16 \ \mu F$

28. Now let us continue with the design procedure.

(6) Design an amplifier with a common emitter gain of about 20. Choose the collector DC voltage to be about half the supply voltage. The main point here is that the collector resistor R_C should be about 1/10 the value of the impedance of the tuned circuit at the resonant frequency. This is often a difficult choice to make, especially if the coil value is not known. Usually an assumption is made, and R_C is chosen arbitrarily.

(7) Stop and connect the circuit shown in frame 20.

(8) Calculate the value of C_C. Do this by choosing X_C to be 160 ohms at the desired frequency. This again is one of those "rules of thumb" which happen to work. It can be justified mathematically. Use the following formula.

$$C_C = \frac{10^6}{2\pi f_r 160} \ \mu F$$

Before continuing substitute the values given so far into the formula to calculate C_C. _____

— — — — — — — — — — — — — —

$$C_C = \frac{10^6}{2\pi \times f_r \times 160\Omega} = 0.1 \ \mu F$$

29. Now, one more step.

(9) Calculate the value of C_B. Again choose it so that X_C is 160 ohms at the desired frequency.

What is the value of C_B? _____

— — — — — — — — — — — — — —

$C_B = 0.1 \ \mu F$

30. The design procedure continues.

(10) Apply power to the circuit and check for oscillations with the oscilloscope. Check the frequency. If it is very far from the desired frequency then adjust C_1 until the desired frequency is obtained. Change C_2 to keep the ratio of values about the same. C_2 will affect the output level.

(11) If the circuit does not oscillate go through the steps outlined in the troubleshooting checklist on this page.

(12) Now alter the feedback connections to the base instead of the emitter. (See frame 22.)

(13) Calculate C_E. X_C should be 160 ohms at the desired frequency. If R_E is less than 1 kΩ, then X_C should be less than 100 ohms.

OSCILLATOR TROUBLESHOOTING CHECKLIST

If an oscillator does not work, most often the trouble is with the feedback connections and a little experimenting—as in steps 2 to 6 below—will produce results. This is especially true when an unknown coil is used which may have several taps or windings. However, each of the following steps should be taken.

(1) Ensure that C_B, C_C, and C_E are all large enough to have a reactance value less than 160 ohms. Ensure that C_E is less than one tenth of R_E.

(2) Check the C_1/C_2 ratio. It should be between 3:1 and 10:1.

(3) Interchange C_1 and C_2; they may be in the wrong place.

(4) Check that the feedback connections are made to the correct place.

(5) Check that the feedback connection is from the correct place.

(6) Check both ends of the tuned circuit; see that they are connected to the correct place.

(7) Check the DC level of the collector, base, and emitter.

(8) Check the capacitor values of the tuned circuit. If necessary change them at random until the circuit oscillates.

(9) If none of the above produce oscillations, check to see if the components are defective. The coil may be opened or shorted. The capacitor may be shorted. The transistor may be dead or its β too low. Check the circuit wiring carefully.

In most cases, one or more of these steps will produce oscillations.

When an oscillator is working it may have one or two main faults.

A distorted output waveform. This is usually caused by C_B, C_C, or C_E not being low enough in value. Other causes are not working in the best frequency range of the coil, or an output amplitude which is too high.

Output level too low. When this happens the sine wave is usually very "clean" and "pure." The best thing is to use another transistor as an amplifier after the oscillator. In a Colpitts oscillator, changing the ratio of C_1 and C_2 will often help.

31.   We will now work through a design example. We will design and build an oscillator with a frequency of 25 kHz, using a coil of 4 mH inductance, using the steps of the procedure shown in frames 27–30.

(1) The value of f_r is given as 25 kHz.

(2) L is given as 4 mH.

(3) Use the formula to find C_1. $C_1 = $ _____

(4) We do not need this step.

(5) Choose C_2. $C_2 = $ _____

(6) We will not go through the amplifier design here. The procedure is shown in Chapter 8.

(7) The circuit will be shown in the next frame.

(8) Find C_C. $C_C =$ _____

(9) Find C_B. $C_B =$ _____

— — — — — — — — — — — — — — —

$C_1 = 0.01\ \mu F$
$C_2 = 0.1\ \mu F$
$C_C = 0.047\ \mu F$ (use $0.1\ \mu F$)
$C_B = 0.047\ \mu F$ (use $0.1\ \mu F$)

Steps 10–13 are the checkout procedure to ensure that the oscillator works. If you build this circuit in a lab, go through steps 10–13. They are not required if the circuit is not built.

32. The circuit designed in frame 31 is shown here.

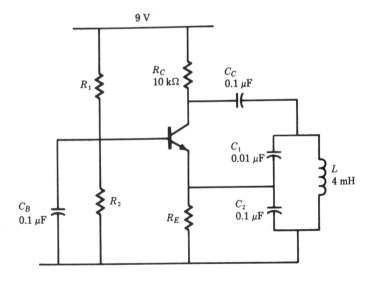

Actual measurements confirmed a frequency very close to 25 kHz. The inductor has a DC resistance of approximately 12 ohms.
Find the approximate impedance of the tuned circuit at resonance.

_____

— — — — — — — — — — — — — — —

$$Z = \frac{L}{C \times r} = \frac{4 \times 10^{-3}}{0.01 \times 10^{-6} \times 12} = 33\ k\Omega \text{ (approximately)}$$

Note this is about three times the value used for R_C.

33. You may wish to work through a second oscillator design example. Assume you need a frequency of 250 kHz using a coil of 500 μH inductance.

   (1) $f_r = 250$ kHz

   (2) $L = 500$ μH $= 0.5$ mH

   (3) Find C_1. $\qquad\qquad\qquad\qquad$ $C_1 =$ _____

   (4) We do not need this step.

   (5) Find C_2. $\qquad\qquad\qquad\qquad$ $C_2 =$ _____

   (6) Use the same amplifier as in the last example.

   (7) The circuit is shown in the next frame.

   (8) Find C_C. $\qquad\qquad\qquad\qquad$ $C_C =$ _____

   (9) Find C_B. $\qquad\qquad\qquad\qquad$ $C_B =$ _____

   $C_1 = 0.0008$ μF
   Using $C_1 = 0.001$ μF and $C_2 = 0.0047$ μF, which are standard values, will produce a C_T value equal to the C_1 found in step 3.
   $C_B = C_C = 0.004$ μF (minimum)

34. The circuit is shown below.

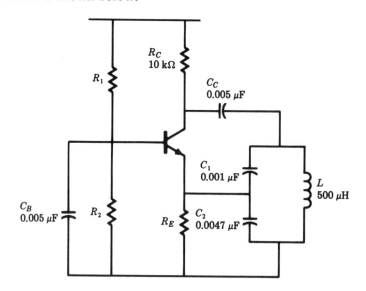

   Measurements showed a frequency close to 250 kHz. The inductor has a DC resistance of approximately 20 ohms.

Find the impedance of the tuned circuit. _____

_ _ _ _ _ _ _ _ _ _ _ _ _ _ _ _ _ _

$Z = 30\ k\Omega$ (approximately)

This is about 3 times the value of R_C.

35. Several other oscillator circuits are shown in this frame. Build as many as you can, and check the measured frequency against the calculated values in each case. Calculate the expected output frequency for each.

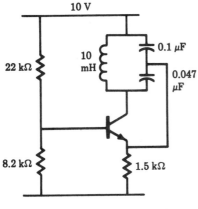

(a) _____

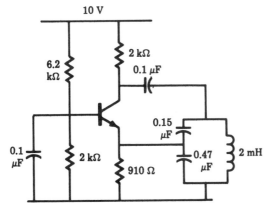

(b) _____

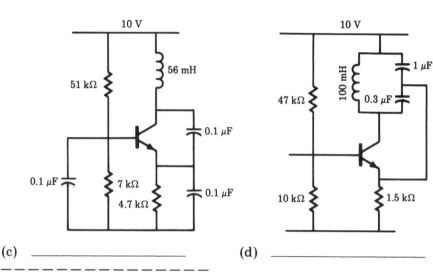

(c) _____   (d) _____

– – – – – – – – – – – – – –

(a) 8.8 kHz
(b) 10 kHz
(c) 3 kHz
(d) 1 kHz

SELF-TEST

The questions below will test your understanding of this chapter. Use a separate.sheet of paper for your diagrams or calculations. Compare your answers with the answers provided following the test.

1. What are the three necessary sections of an oscillator? _____

_____

2. What is the difference between positive and negative feedback?

_____

_____

3. What type of feedback is needed in an oscillator? _____

4. What is the formula for the frequency of an oscillator?

_____

5. Draw a simple circuit for a Colpitts oscillator.

6. Draw a simple circuit for a Hartley oscillator.

7. Draw a simple circuit for an Armstrong oscillator.

8. Frames 27–30 gave a design procedure for oscillators. How well do the circuits in frame 35 fulfill the criteria given? In other words, check the values of V_f, A_V in CE mode, C_1/C_2 ratio, R_C/Z ratio, and the frequency.

(a) _____

(b) _____

(c) _____

(d) _____

9. Using the basic amplifier of frame 34, design an oscillator to have a frequency of 10 kHz using a 100 mH coil. Calculate the values of C_1, C_2, C_C, and C_B._____

Answers to Self-Test

If your answers do not agree with those given below, review the frames indicated in parentheses before you go on to the next chapter.

1. an amplifier, feedback, and a resonant load   (frame 1)

2. Positive feedback is "in phase" with the input, and negative feedback is "out of phase" with the input.   (frames 2–3)

3. positive feedback   (frame 3)

4. $f_r = \dfrac{1}{2\pi\sqrt{LC}}$ (frame 11)

5. See frame 20.

6. See frame 25.

7. See text at end of frame 25.

8. (a) $V_f = \dfrac{0.047}{0.147}$

   A_V cannot be calculated
   $C_1/C_2 = 0.047/0.1 = 0.47$
   Z cannot be calculated since r is unknown
   $f_r = 8.8$ kHz (approximately)

   (b) $V_f = \dfrac{0.15}{0.62}$

   $A_V = \dfrac{2000}{910} = 2.2$ (approximately)

   $C_1/C_2 = 1/3$ (approximately)
   Z cannot be calculated
   $f_r = 10$ kHz (approximately)

(c) $V_f = \dfrac{0.1}{0.2}$

A_V cannot be calculated
$C_1/C_2 = 1$
Z cannot be calculated
$f_r = 3$ kHz

(d) $V_f = \dfrac{0.3}{1}$

A_V cannot be calculated
$C_1/C_2 = 0.3$
Z cannot be calculated
$f_r = 1$ kHz (approximately)   (frames 27–30)

9. $C_1 = 0.0033 \ \mu F$; $C_2 = 0.01 \ \mu F$; $C_B = C_C = 0.1 \ \mu F$ (frames 26–30)

CHAPTER TEN

The Transformer

Transformers are used to "transform" an AC voltage to a higher or lower level. For example, in a TV set one transformer produces 28 V from the 120 V power line; and another produces several thousand volts from a 28 V oscillator.

Transformers can be seen near the top of power line poles in the streets and as large fenced off installations in factories. They are also found in communications equipment and digital equipment, where often they are the size of a fingernail.

The communications uses of transformers are usually to "isolate" items of equipment to eliminate electrical noise and other static interference.

When you complete this chapter you will be able to:

- recognize a transformer in a circuit;

- explain and correctly apply the concepts of "turns ratio" and "matching";

- recognize two types of transformer;

- do simple calculations involving transformers.

TRANSFORMER BASICS

1. Consider two coils placed very close to each other. If an AC voltage is applied to one of them causing a current to flow, it will set up a changing (or alternating) magnetic field around the coil. Because of the close proximity of the second coil to the first coil, much of the changing magnetic flux cuts the turns of the second coil. This will cause an AC voltage to be induced across the terminals of the second coil. The voltage output of the second coil, which will be at the same frequency as the input voltage, can now be used to supply the needs of an external circuit.

Coil 1   Coil 2

The configuration of the two coils and the manner in which they are coupled together (magnetically) creates a device that falls in the realm of magnetic circuits. Any study of basic electricity should also include a review of electromagnetism, which involves concepts for magnetic circuits. The purpose of this chapter is to view this particular magnetic circuit device, called a transformer, as a basic component in an electronic circuit. A more detailed study of magnetic circuits and transformers is left to other books.

The inductive coupling of two coils is improved the closer the two coils are wound together. Often coils are wound on a common iron core called a "former," which gives maximum coupling since magnetic flux exists or flows easily in the iron. Also, the design of the transformer is not limited to two coils but can have numerous output coils.

(a)   When the two coils are wound together are they connected electrically? _____

(b)   What type of device is described by the coupling effect of two or more coils? _____

(c)   What is the name given to this device? _____

(d)   What would be the effect on all the other coils if an AC voltage were applied to the terminals of the first coil? _____

_____

_____

— — — — — — — — — — — — — — — — —

(a)   no
(b)   a magnetic circuit device
(c)   transformer
(d)   An AC voltage of the same frequency would appear at the terminals of each of the other coils.

Note: Other devices that are actually based on magnetic circuit principles include such common items as the relay, the speaker, and the microphone.

2.   A transformer is an AC device only. A DC voltage applied to one set of terminals will have no effect on the other terminals.

However when a sine wave signal is applied to one coil, a sine wave of the same frequency will be observed across the other coil.

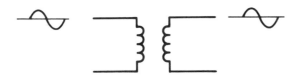

The coil to which the signal is applied is usually called the *primary* coil. The induced voltage appears across the other coil, which is normally called the *secondary* coil.

(a) What will be the difference in frequency between the signal at the primary coil and that at the secondary? _____

(b) What will be the output at the secondary coil if 10 V DC is applied to the primary coil? _____

- - - - - - - - - - - - - - - - - -

(a) no difference—the frequencies will be the same
(b) DC does not produce the changing magnetic field that is necessary to induce a voltage output at the second coil. Hence, there will be no output at the secondary coil. Note: We often summarize this by saying that DC does not pass through a transformer.

3. If one side of each coil is grounded, as shown in the diagram, then the input and output waveforms can be compared. If the output goes positive when the input goes positive, as in the figure, then they are said to be in phase.

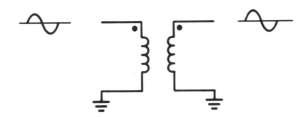

The dots in the diagram show which ends of the coils produce in-phase voltages. If one coil is reversed then the output will be inverted from the input. It is said to be out of phase and a dot is placed at the opposite end of the coil.

In the diagram on the next page, the primary coil on the left has the same phase as before. Put a dot in the secondary coil to show that it is out of phase.

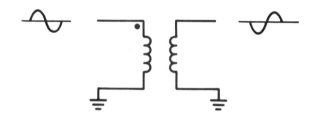

The dot should be at the lower end of the right coil.

4. The two types of secondary winding on a transformer are shown below.

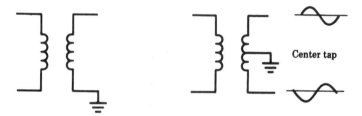

Center tap

When a transformer has no center tap, one side of the secondary is often
(but not always) grounded in a circuit. With a center tap, the tap is
often grounded.

From the above diagrams, what can you tell about the output from

the two ends of the center-tapped secondary? _____

- - - - - - - - - - - - - - - - - -

They are out of phase.

5. All practice and theory about transformers is based on this one fact.

Power in = Power out,   or $P_{in} = P_{out}$

In practice, transformers are imperfect. Some slight loss of power
occurs, which is measured by the *efficiency* of the transformer. Here
we will assume the transformer is perfect—it has an efficiency of 100%.

If the number of turns of wire in the secondary is increased, a
larger output voltage will be induced in the secondary. In a transformer
the output voltage from the secondary will be directly proportional to
the number of turns in the secondary.

How does increasing the number of turns of wire in a secondary

coil affect the output voltage in the secondary? _____

_____

- - - - - - - - - - - - - - - - -

It increases the output voltage from the secondary.

6. The transformer below shows the number of turns in the primary and secondary coils as N_p and N_s, respectively.

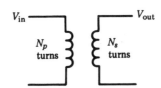

V_{in} V_{out}

N_p turns N_s turns

The ratio of the input to output voltage is the same as the ratio of the primary to secondary turns. Write a simple formula to express this.

− − − − − − − − − − − − − − − −

$$\frac{V_{in}}{V_{out}} = \frac{N_p}{N_s}$$

Note: The ratio of primary turns to secondary turns is called the turns ratio (TR).

7. Use the formula you just wrote down to calculate the output voltage when a 10 Vpp sine wave is applied to a transformer with a 2 to 1 (2:1) turns ratio. _____

_____

− − − − − − − − − − − − − − − −

$$\frac{V_{in}}{V_{out}} = \frac{N_p}{N_s} = TR$$

$$\frac{V_{in}}{V_{out}} = \frac{N_p}{N_s} = \frac{2}{1} = TR$$

$$V_{out} = V_{in} \times \frac{1}{TR} = 10 \times \frac{1}{2} = 5 \; V_{pp}$$

8. Now try these few examples.

(a) V_{in} = 20 Vpp, turns ratio = 5:1. V_{out} = _____

(b) V_{in} = 1 Vpp, turns ratio = 1:10. V_{out} = _____

(c) V_{in} = 100 Vrms. Find V_{out} when an equal number of turns are on the primary and the secondary. V_{out} = _____

− − − − − − − − − − − − − − − −

(a)   4 Vpp (This is sometimes called a "step down" transformer.)
(b)   10 Vpp (This is sometimes called a "step up" transformer.)

(c) 100 V$_{rms}$ (This is sometimes called an isolation transformer, which is used to separate or isolate the voltage source and the load electrically.)

9. Almost all electronic equipment which is operated from the 120 V AC house current requires a transformer to convert the 120 V AC to a more suitable lower voltage. A typical example is the 28 V secondary winding found on many transformers.

120 V AC   28 V AC

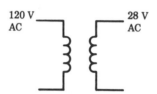

Calculate the turns ratio for this transformer. _____

- - - - - - - - - - - - - - - -

$$\frac{N_p}{N_s} = \frac{120}{28} = 4.3{:}1 \text{ (approximately)}$$

10. The output of the 28 V secondary, when viewed on an oscilloscope, looks like the following diagram.

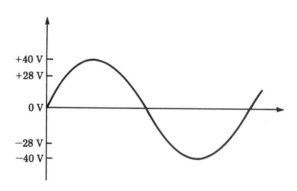

(a) Is 28 V a peak-to-peak or an rms value? _____

(b) What is the peak-to-peak value of the 28 V at the secondary?

_____

- - - - - - - - - - - - - - - -

(a) rms

(b) 2 × 1.414 × 28 = 79.184 V

11. Like the 28 V, the 120 V wall plug value is an rms measure. What is the peak-to-peak value of the voltage from the wall plug? _____

_ _ _ _ _ _ _ _ _ _ _ _ _ _ _ _ _

approximately 340 volts

12. The actual voltage measured at the secondary of a transformer depends upon where and how the measurement is made. Look at the secondary below and notice how the measurements are made.

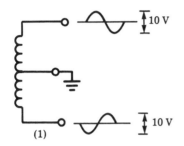 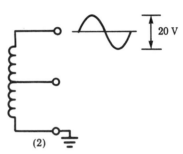

If the center tap is used and grounded (as shown in diagram 1), then 10 V_{pp} of opposite phase will be seen at each terminal. If the center tap is not used (as in diagram 2), but one end is grounded, then a 20 V_{pp} voltage will be seen at the other terminal.

(a) Assume a center-tapped secondary is rated at 28 V referenced to the center tap. What is the rms voltage output when the center tap is used? _____

(b) Assume the 28 V is the total output voltage across the entire secondary winding. What will be the output at each end?

_____

(c) Assume a center-tapped secondary has 15 V output. What is the pp output when the center tap is not used? _____

_ _ _ _ _ _ _ _ _ _ _ _ _ _ _ _ _

(a) 28 V_{rms} at each end
(b) 14 V (one half of the total V_{out})
(c) When the center tap is used, the output is 15 V_{rms}. So when it is not used, the output is 30 V_{rms}.

V_{pp} = 2 × 1.414 × 30 = 84.84 volts

13. It is useful to note the current relationship for a transformer. From the "*Power in = Power out*" expression of frame 5, we get

$$V_{in}I_{in} = V_{out}I_{out}$$

which yields

$$\frac{V_{in}}{V_{out}} = \frac{I_{out}}{I_{in}} = TR$$

This means that voltage and current are inversely related. This relationship would have to be so in order to keep the powers equal. Thus, we see that the output current equals the input current times the turns ratio, whereas the output voltage equals the input voltage divided by the turns ratio.

(a) What would be the input current for a transformer if the input power was 12 watts at a voltage of 120 V_{rms}? _____

_____

(b) What would be the transformer's output voltage if the turns ratio was 5:1? _____

_____

(c) What would be the output current? _____

_____

(d) What would be the output power? _____

_____

— — — — — — — — — — — — — — — —

Note: In AC power calculations, the RMS values of current and voltage must be used.

(a) $I_{in} = \dfrac{P_{in}}{V_{in}} = \dfrac{12}{120} = 0.1$ A$_{rms}$

(b) $V_{out} = \dfrac{V_{in}}{TR} = \dfrac{120}{5} = 24$ V$_{rms}$

(c) $I_{out} = I_{in} (TR) = 0.1 \times 5 = 0.5$ A$_{rms}$

(d) $P_{out} = V_{out}I_{out} = 24 \times 0.5 = 12$ watts; same as the power in

TRANSFORMERS IN COMMUNICATIONS CIRCUITS

14. In communications circuits an input signal is often received via a very long interconnecting wire, usually called a *line*, which normally has an impedance of 600 ohms. A typical example is a telephone line between two cities. The output of the sending equipment is also connected to the line.

Communications equipment works best when connected to a load which has the same impedance as the output of the equipment. What output impedance should communications equipment have? _____

— — — — — — — — — — — — — — — — —

600 ohms output impedance, to be connected to a 600 ohms line

15. Since most electronic equipment does not have a 600 ohm output impedance, a transformer is often used to connect such equipment to a line. Often the transformer will be built into the equipment for convenience. The transformer is used to "match" the equipment to the line.

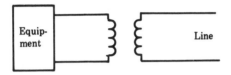

To work correctly the output of the transformer secondary should have a 600 ohm impedance to exactly match the line. The output impedance of the transformer, measured at the secondary winding, is governed by two things. One is the output impedance of the equipment.

What would you expect the other to be? _____

— — — — — — — — — — — — — — — — —

the turns ratio of the transformer (The DC resistance of the wire has no effect and can be ignored.)

16. Assume a signal generator with an output impedance of Z_G is connected to one side of a transformer. In the diagram, the impedance is shown in parallel. A load impedance of Z_L is connected to the other side. Let us see what effect the transformer has by looking at the input and output powers in terms of the external impedances. We will be using Power = V^2/Z.

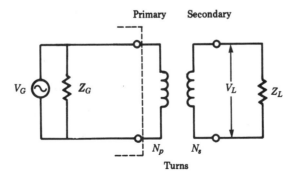

$$P_{in} = P_{out}$$

which means

$$\frac{V_G^2}{Z_G} = \frac{V_L^2}{Z_L}$$

$$\frac{Z_G}{Z_L} = \left(\frac{V_G}{V_L}\right)^2$$

Since V_G and V_L are directly across the transformer primary and secondary,

$$\frac{Z_G}{Z_L} = \left(\frac{N_p}{N_s}\right)^2 = (TR)^2$$

Thus, we see that the ratio of the impedances on either side of the transformer is related as the square of the turns ratio. This relationship will allow us to use the transformer to match impedances in such a way that the generator will see an impedance equal to its own impedance. The load will also see an impedance equal to itself. A particular turns ratio is required to perform the matching.

Now work the following problem using the data given. A generator has an output impedance of 10 kΩ and it produces a 10 V_{pp} signal. It is to be connected to a 600 ohm line. The frequency of the generator does not enter into the calculations; hence, matching will work at all frequencies. However, since there is a frequency range for all transformers, you must not use a transformer at a frequency outside of the range it is designed for.

(a)   To properly match the generator to the line, what turns ratio is required? _____

_____

(b)   Find the output voltage across the load. _____

_____

(c)   Find the load current and power. _____

_____

_____

- - - - - - - - - - - - - - - -

(a)   4.08:1

(b) $V_L = \dfrac{V_G}{TR} = \dfrac{10}{4.08} = 2.45$ V_{pp}, which is 0.866 V_{rms}

(c) $P_{in} = \dfrac{V_G^2}{Z_G} = \dfrac{(3.53)^2}{10,000} = 1.25$ mW

Note: For the power calculation, the RMS value of the voltage must be used.

$$I_{in} = \frac{1.25 \text{ mW}}{3.53 \text{ V}} = 0.354 \text{ mA}_{rms} = 1 \text{ mA}_{pp}$$

$$I_L = I_{in}(TR) = 1.445 \text{ mA}_{rms} = 4.08 \text{ mA}_{pp}$$

$$P_L = \frac{(2.45)^2}{600} = (0.001445)^2 600 = 1.25 \text{ mW; the same as the}$$

input power

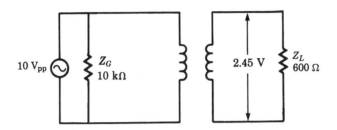

Note: The generator now sees 10 kΩ when it looks toward the load rather that the actual 600 ohm load. By the same token, the load now sees 600 ohms when it looks toward the source. This condition allows the best transfer of power to take place between the source and the load. In practice, however, the optimum condition as calculated above rarely exists. As a turns ratio of 4.08:1 may not be obtainable, you would have to select the closest value, which may be a turns ratio of 4:1. This will affect the conditions at the load side, but only slightly.

17. Now work this example:

    (a)   Match a generator that has a 2 kΩ output to a 600 ohm line. What transformer ratio is required? _____

    _____

    (b)   If the generator produces 1 V$_{pp}$, what is the voltage at the line?

    _____

– – – – – – – – – – – – – – – – – –

    (a)   TR = 1.83
    (b) $V_L = 0.55 \text{ V}_{pp}$

18. Now work this problem:

    (a)   What is the turns ratio required to match a 2 kΩ load with a source that has an output impedance of 5 kΩ? _____

(b) If the load requires a power of 20 mW, what should the source voltage be set at? (First find the voltage across the load.) _____

_____

_____

(c) What will be the primary and secondary currents and the power supplied by the source to the primary side of the transformer? ____

_____

_____

– – – – – – – – – – – – – – – – –

(a) TR = 1.58
(b) $V_L = 6.32$ V$_{rms}$, $V_G = 10$ V$_{rms}$
(c) $I_L = 3.16$ mA$_{rms}$, $I_p = 2$ mA$_{rms}$, $P_{in} = 20$ mW

SUMMARY AND APPLICATIONS

In practice, transformers are made with a given turns ratio, and often the impedances between which they must be connected are specified. For example, a transformer will be marked "Pri: 10 kΩ, Sec: 2 kΩ." The turns ratio will not be quoted, since you will not need to use it.

Transformers are a very complex part of electronics and have been covered very briefly here. They are made in many different sizes and usually to work over a specific frequency range. Those used for audio work are quite unsuitable for transforming the 120 V wall power to a lower voltage, and equally different are the transformers sometimes found in the antenna lead of a TV set. But in all cases, the basic rules of power and turns ratio apply. Some of the main uses of transformers are explained briefly below.

(1) The 120 V AC from the wall plug can be changed to a lower AC voltage. This is then converted to a DC voltage in a power supply circuit. This is how much electronic equipment is powered.

(2) AC communication signals—such as voice, music, or TV—from a distant source can be connected to the input of local equipment. This is very common in broadcasting and telephones.

(3) A similar use is to connect the output of local communications equipment to the lines between cities so that messages may be sent across them.

(4) In the above use, the impedance as well as the voltage level of the signal can be changed by the transformer. Either maximum power or maximum voltage is transferred from one piece of equipment to the other.

(5) The various stages in amplifiers are often coupled together with transformers. In audio amplifiers, the output is often coupled to the speaker with a transformer.

SELF-TEST

The questions below will test your understanding of this chapter. Use a separate sheet of paper for your diagrams or calculations. Compare your answers with the answers provided following the test.

1. How is a transformer constructed? _____

    _____

2. What is used as an input to a transformer? _____

3. If a sine wave is fed into the transformer diagrammed here, what is the

    output? _____

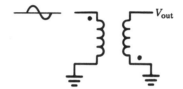

4. What is meant by the term *turns ratio?* _____

5. If V_{in} = 1 V_{pp} and TR = 2, what is the output voltage?

    V_{out} = _____

6. If V_{in} = 10 V_{pp} and V_{out} = 7 V_{pp}, what is the turns ratio?

    TR = _____

7. In this center-tapped secondary winding, the voltage between points A and B may be expressed as V_{A-B} = 28 V_{pp}. What is the voltage between

    C and A? _____

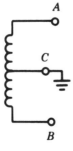

8. In the same secondary, the voltage between points B and C is $V_{B-C} =$ 5 V_{rms}. What is the peak-to-peak voltage between A and B? _____

_____

9. If $I_{in} = 0.5$ A_{rms} and $I_{out} = 2.0$ A_{rms}, what is the turns ratio? _____

10. Is the transformer of problem 9 a step up or a step down transformer?

_____

11. If $Z_L = 600$ ohms and $Z_G = 6$ kΩ, find the turns ratio.

TR = _____

12. If $Z_L = 1$ kΩ and the turns ratio is 10:1, what is the generator impedance?

$Z_G =$ _____

Answers to Self-Test

If your answers do not agree with those given below, review the frames indicated in parentheses before you go on to the next chapter.

1. two coils of wire wound around a common iron former (frame 1)
2. an AC voltage—DC will not work (frame 2)
3. an inverted sine wave (frame 3)
4. the ratio of the turns in the primary winding to the number of turns in the secondary (frame 6)
5. $V_{out} = 0.5$ V (frame 7)
6. TR = 1.43:1 (frame 7)
7. $V_{C-A} = 14$ V_{pp} (frame 12)
8. $V_{A-B} = 14.14$ V_{pp} (frame 12)
9. TR = 4:1 (frame 13)
10. It is a step down transformer. The voltage will step down when the current is stepped up. This will maintain the same power on either side of the transformer. (frame 13)
11. TR = 3.2:1 (frame 16)
12. $Z_G = 100$ kΩ (frame 16)

CHAPTER ELEVEN

AC Diode Circuits
and Power Supplies

A power supply is incorporated into most electronic equipment. It takes the 120 V AC from a wall plug and converts it to a DC voltage that is used to provide power for the electronic circuits.

Power supply circuits are very simple in principle, and those shown here are basically the same types which have been used for many years. Power supplies incorporate many of the features covered earlier in this book and make an excellent conclusion to your study of basic electronics.

Diodes are a major component in power supplies, and how diodes behave in the presence of an AC signal is a fundamental part of how power supplies work. So this chapter begins with a brief discussion of diodes in AC circuits.

When you complete this chapter you will be able to:

- describe the effect of diodes in AC circuits;

- identify at least two ways of rectifying an AC signal;

- calculate the output voltage from a rectifier;

- determine the appropriate components for a power supply circuit;

- draw the output waveform from a rectifier and a smoothing circuit.

DIODES IN AC CIRCUITS

1. Diodes can be used for several different purposes in AC circuits, where their characteristic of conducting in only one direction is necessary. Before seeing how diodes behave in the presence of an AC signal, let's think back to Chapter 2 for a moment. Look at the circuit at the top of the next page, and assume 20 V DC is applied at point A.

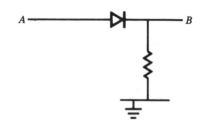

(a) What is the output voltage at point B? _____

(b) Suppose a 10 V DC is now applied at B. What is the voltage at
       point A? _____

- - - - - - - - - - - - - - - -

(a) 20 V DC (Ignore for now the voltage drop of 0.7 V across the
       diode.)
(b) 0 V

2. Suppose now that a 20 V_{pp} AC signal is superimposed upon the 20 V
     DC level.

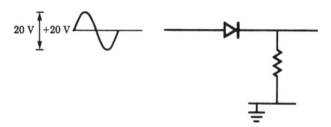

(a) What are the positive and negative peak voltages of the input
       circuit? _____

(b) What is the output voltage in this case? _____

- - - - - - - - - - - - - - - -

(a) 30 V and 10 V
(b) The diode is always forward biased, so it always conducts. Thus
       the output is exactly the same as the input.

3. Suppose now we have the 20 V_{pp} AC signal, but it is centered around
     0 V DC, or ground. As before, when the input is positive the diode will
     conduct.

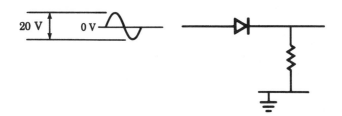

(a) What are the positive and negative peak voltages of the input signal? _____

(b) For the positive half wave of the input, draw the output.

_ _ _ _ _ _ _ _ _ _ _ _ _ _ _ _ _ _

(a) +10 V and −10 V

(b)

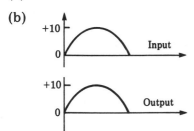

4. When the input is negative, the diode is nonconducting; thus the output voltage remains at 0 V.

The input waveform is shown below. Draw the output waveform underneath it.

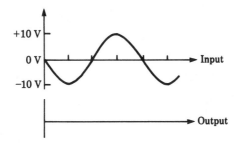

_ _ _ _ _ _ _ _ _ _ _ _ _ _ _ _ _ _

5. Now you know what the output looks like for one complete cycle of the input waveform. Draw the output for three complete cycles.

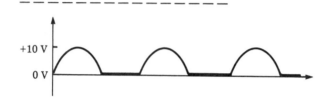

6. If the diode is reversed, it will conduct on the negative parts of the cycle and be reverse biased on the positive parts of the cycle. This will invert the output waveform seen in frames 4 and 5.

    Draw the output waveform for the three input cycles, assuming a reversed diode from that used in frame 5.

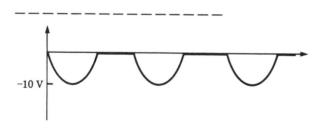

7. Consider now a 20 V_{pp} AC signal superimposed on a -20 V DC level, as in the following figure.

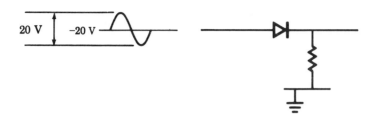

(a)   When is the diode forward biased? _____

(b)   What will the output voltage look like? _____

- - - - - - - - - - - - - - - - -

(a)   never, as the signal remains negative
(b)   0 V all the time

8.   As you have seen, a diode limits an AC signal to only one side of the
zero axis. This is called *rectifying* the AC. An AC signal must always be
rectified before it can be converted to a DC signal.
   Refer back to the rectified signals in frames 5 and 6. Would you

expect these to be converted to positive or negative DC voltages? _____

_____

- - - - - - - - - - - - - - - - -

in frame 5, to a positive voltage; in frame 6, to a negative

9.   Diodes are always used in electronic power supplies. The diode can be
connected to the secondary coil of a transformer as shown below.

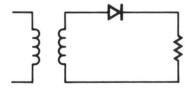

(a)   What will be the effect of the diode on the AC signal? _____

(b)   What would the waveform of the voltage across the load look like
   if the secondary of the transformer is a 30 V_{pp} AC signal centered

   around 0 V? _____

- - - - - - - - - - - - - - - - -

(a)   It will become rectified.
(b)   It will look like this. This is called *half-wave rectification*.

If you have the appropriate facilities, you can set up this circuit and observe the half-wave rectification effect. Reverse thé diode, and see the negative half waves on the oscilloscope as well.

10. If the transformer secondary has a center tap, this rectifier circuit can be used. Examine the circuit below.

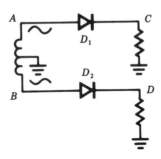

(a) Which diode will conduct on the first half wave? _____

(b) Which diode will conduct on the second half wave? _____

(c) Draw the input waveforms from point A and point B to D_1 and D_2, and underneath each draw its output waveform (points C and D).

— — — — — — — — — — — — — — — — —

(a) On the first half wave, D_1 conducts and D_2 is reverse biased.
(b) On the second half wave, D_2 conducts and D_1 is reversed.
(c)

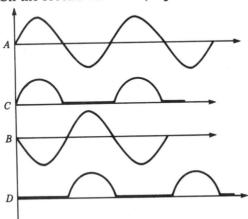

11. If the two diodes in the rectifier circuit of frame 10 are connected across the same resistor, as on the following diagram, then their output waveforms will be combined across this common load.

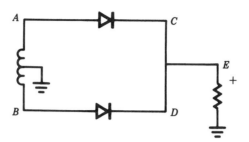

Refer to frame 10 and draw the combined output waveform at point *E.*

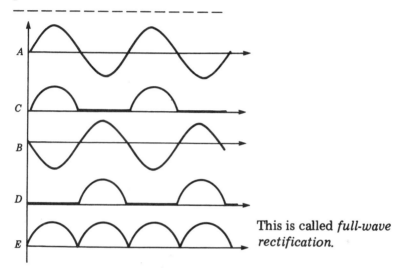

This is called *full-wave rectification.*

Note: The frequency of the rectified output waveform is now 120 Hz, whereas the frequency of the half-wave output was 60 Hz.

If you have the facilities, you will want to set up this full-wave rectified circuit and observe the output on the oscilloscope.

12. Full-wave rectification of AC allows a much "smoother" conversion of the AC to DC than does half-wave rectification. We have seen a full-wave rectifier built from a transformer with a center-tapped secondary.

But a full-wave rectifier can also be built using a transformer without a center-tapped secondary, as shown here.

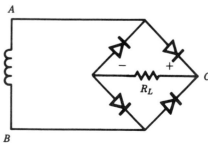

How does this circuit differ from the previous circuit? _____

_____

– – – – – – – – – – – – – – – –

It has no center tap on the secondary, and it has four diodes.

13. When point A goes positive, the conduction path is as shown.

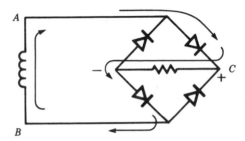

On the other hand, when point B goes positive, this is the conduction path.

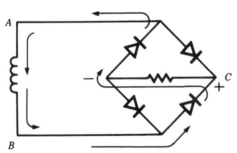

Notice that the direction of current through the load is the same in both cases.

(a) Through how many diodes does the current travel in each conduction path? _____

(b) Draw the voltage waveform at point C.

– – – – – – – – – – – – – – –

(a) two diodes in each case

(b)

For those who have facilities, this rectifier should also be set up and observed on an oscilloscope.

A rectified AC is often called a "pulsating DC." It is most commonly found in power supply circuits, which we will study next.

If you want to take a break, this is a good stopping point.

POWER SUPPLIES

14. The basic power supply circuit can be divided into four sections.

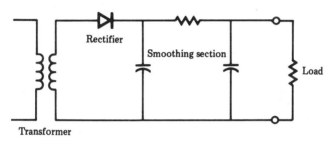

There are several variations to be found within each section, but we will only look at the basic circuits.

The transformer was covered in the last chapter. In the power supply we use a transformer which converts the 120 V AC to a lower AC voltage. The choice of this AC depends on the final level of DC required.

Recall our rectifier discussion.

(a) What number of diodes would produce a full-wave rectified output if a center-tapped transformer is used? _____

(b) In the above diagram, will full- or half-wave rectification result?

_____

(c) What will be the output of the rectifier section? _____

- - - - - - - - - - - - - - -

(a) two
(b) half-wave
(c) pulsating DC

15. The function of the smoothing section is to take the pulsating DC and convert it to a "pure" DC with as little AC "ripple" as possible. The smoothed DC is then applied to the load.

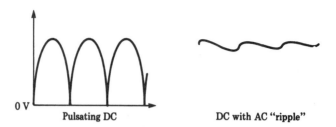

Pulsating DC   DC with AC "ripple"

The load which is "driven" by the power supply can be a simple lamp or a complex electronic circuit. Whatever is used, it will require a certain voltage across its terminals and will draw a certain current. So obviously it will have a resistance.

Usually the voltage and current required—and hence the resistance—are known, and the power supply must be designed to provide the voltage and current.

We can treat the load as a simple resistor to keep our diagrams uncomplicated.

(a)   What does the smoothing section of a power supply do? _____

_____

(b)   What is connected to a power supply, and what can we treat it

like? _____

_____

— — — — — — — — — — — — — — — —

(a)   It converts the pulsating DC to a "pure" DC.
(b)   A load such as a lamp or an electronic circuit is connected to a power supply. In most cases we can treat it like a resistor.

16.   Below is a simple power supply circuit, with a resistor as the load.

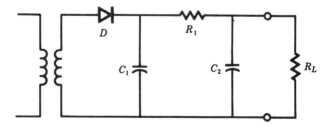

(a)   What type of transformer is used? _____

(b)   What type of rectifier is used? _____

(c)   What components make up the smoothing section? _____

_____

(d)   What output would you expect from the rectifier section?

_____

— — — — — — — — — — — — — — — —

(a)   a secondary with no center tap
(b)   a single diode half-wave rectifier
(c)   a resistor and two capacitors (R_1, C_1, and C_2)
(d)   half-wave pulsating DC

17.   The output from the rectifier is this waveform.

This is applied to the smoothing section and the load. Consider the first half wave only.

As this voltage level rises to its peak value, it charges capacitor C_1 to this peak value.

When the input drops back to 0 V, we have the situation shown below.

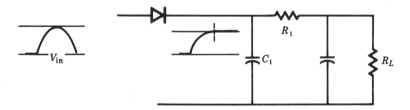

What discharge path is available for the capacitor C_1? _____

_____

— — — — — — — — — — — — — — — —

The diode is now reverse biased, so the capacitor cannot discharge through the diode. Its only discharge path is through R_1 and the load R_L.

18. The value of C_1 is chosen so that the discharge time constant is long. If no further half waves arrive at the input, the waveform would look like that shown below.

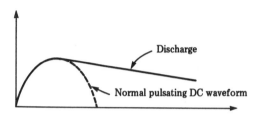

However, if another half wave arrives, the input and output waveforms would look like this.

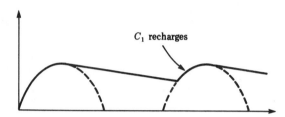

The capacitor only discharges a small amount before it is recharged to the peak value.

Further half waves will repeatedly produce this same recharging effect. The resulting waveform at C_1 will thus be as shown below.

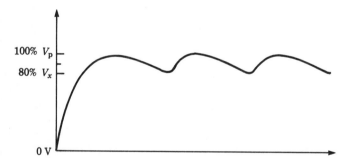

This is, in effect, a DC level which varies between V_p and V_x. If the discharge time constant of C_1 through R_1 and R_L is made about 10 times the duration of the input pulses, then V_x will be about 80% of V_p.

If the discharge time selected is greater than 10 times the duration of the input pulses, the smoothing will be even better. For design purposes, however, we will select a value of 10 times the duration throughout this chapter, which will yield practical design values and allow us to make easy approximations.

What would you estimate to be the average DC output level of this circuit? _____

— — — — — — — — — — — — — — — — —

about 90% of V_p

19. Assume the transformer has a 28 V_{rms} output from its secondary, and we require 10 V DC across a 100 ohms load. Let us work out the component values needed.

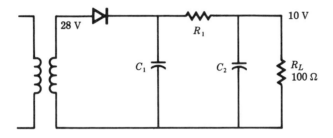

What is the peak voltage out of the rectifier? _____

— — — — — — — — — — — — — — — — —

The transformer secondary delivers 28 V_{rms}, so the peak voltage from the rectifier will be 1.414 times this rms value. This is about 40 V.

20. The input waveform to the smoothing section is shown below.

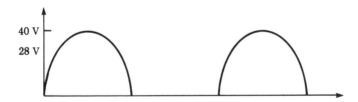

This is a 60 Hz signal. Calculate the duration of a half wave. _____

_____

— — — — — — — — — — — — — — — — —

60 Hz means 60 complete sine waves in 1 second. One sine wave thus lasts for 1/60 second.

1/60 second = 1000/60 milliseconds = 16.67 ms

So the duration of a half wave is 8.33 ms.

21. The average DC level after smoothing will be about 90% of the peak value of the half sine wave (which in our example is 40 V). This is with the approximation suggested in frame 18 regarding the selection of a time constant. R_1 and R_L must act as a voltage divider to reduce the 36 V DC level to the required 10 V DC at the output.

Use the voltage divider formula to find an appropriate value for R_1. R_L has already been given as 100 ohms.

_ _ _ _ _ _ _ _ _ _ _ _ _ _ _ _ _

$$V_{out} = \frac{V_{in} R_L}{(R_1 + R_L)}$$

$$10 = \frac{36 \times 100}{(R_1 + 100)}$$

Thus, $R_1 = 260$ ohms

22. The circuit now has these values.

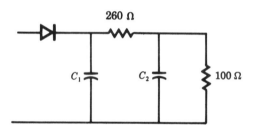

We must now chose C_1 so that its discharge time through the two resistors is 10 times the input wave duration.

(a) Refer back to the preceding frames. How long should the discharge time constant be? _____

(b) From the time constant, find the value of C_1. _____

_____

_ _ _ _ _ _ _ _ _ _ _ _ _ _ _

(a) Since the input duration is 8.33 ms, the time constant should be at least 83.3 ms.

(b) $T = R \times C = (R_1 + R_L) \times C_1$
$0.0833 = 360 \times C_1$ which yields $C_1 = 230$ μF

Note: Capacitance values as large as this will have to be supplied by electrolytic capacitors. A value larger than 230 μF would also be acceptable. A value less than 230 μF would not meet the requirement of frame 18.

23. The circuit and the voltage waveforms at various points are shown below.

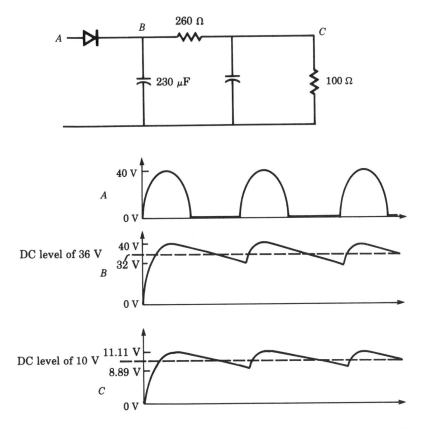

(a) What has happened to the DC output voltage between B and C?

_____

(b) What has happened to the AC between A and C? _____

_____

– – – – – – – – – – – – – – – – – –

(a) It has been divided down by the voltage divider from 36 V to 10 V.
(b) The AC is changed from the half wave to an AC ripple riding on a 10 V DC level.

24. In most cases this DC level across the load still pulsates too much. In other words, the AC ripple is still too high. Further smoothing is required. This is provided by C_2.

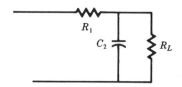

The method of deciding on the value of C_2 is different from that used for C_1. We now have a voltage divider action of R_1 and the parallel combination of R_L and X_{C_2}.

If C_2 is chosen so that its reactance is equal to or less than one tenth of the value of R_L, then the AC voltage divider is approximately given by R_1 and X_{C_2} only. The effect of R_L is negligible. The technique for making calculations with an R,C voltage divider is given in Chapter 6.

(a) What value in ohms do we need for X_{C_2}? _____

(b) What is the formula for the reactance of a capacitor?

_____

(c) What is the frequency in this case? _____

(d) Now find the value of the capacitor C_2. _____

- - - - - - - - - - - - - - - -

(a) not more than one tenth of R_L

(b) $X_C = \dfrac{1}{2\pi fC}$

(c) 60 Hz, as we are dealing with the AC power line

(d) using 10 ohms for X_c and $C_2 = \dfrac{1}{2\pi fX_c}$

$C_2 = 265 \ \mu F$

25. The final circuit is shown below.

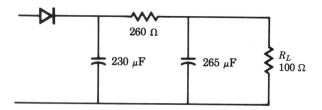

The AC ripple now sees this voltage divider.

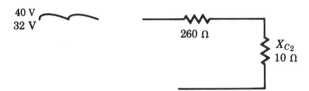

(a)   What is the peak-to-peak voltage at the input to the AC voltage divider? _____

(b)   Find the AC ripple output across R_L using the AC voltage divider. _____

- - - - - - - - - - - - - - - -

(a) $8\ V_{pp}$

(b) $AC\ V_{out} = (AC\ V_{in}) \times \dfrac{X_{C2}}{\sqrt{(X_{C2}^2 + R_1^2)}}$

$AC\ V_{out} = 8 \times \dfrac{10}{\sqrt{(10^2 + 260^2)}} = 0.31\ V_{pp}$

Note: This means that the addition of C_2 will lower the AC ripple shown by curve C of frame 23 (peak values of 11.11 and 8.89) to new values of 10.155 and 9.845 V. This represents a lower ripple at the output, hence C_2 aids the smoothing of the 10 V DC at the output.

26.   This same procedure can be used with a full-wave rectifier. Let us use the same conditions as in the last example. We have a secondary which supplies 28 V_{rms} and we wish to have 10 volts DC across a 100 ohm load.

Shown below is the output waveform from the rectifier.

The peaks are now adjacent, rather than spaced as they were in the previous case, shown in frame 18. When capacitor C_1 is added, a smoothed waveform will result.

If we again choose the discharge time constant to be 10 times as long as the period of the waveform, the V_x will now be about 90% of V_p. The average DC level will be about 95% of V_p.

(a) What was the average DC level in the previous case? _____

(b) What is the DC level with a full-wave rectifier that has the rectifier output waveform as shown above?

_____

(c) Why do you suppose a higher value is obtained than in the previous

case? _____

_____

– – – – – – – – – – – – – – – – –

(a) 36 V which was 90% of V_p
(b) 38 V which is 95% of V_p
(c) The slightly higher values occur because the capacitor does not discharge as far with full-wave rectification, and so a slightly higher DC and slightly less AC ripple are obtained.

27. We can now calculate the value of R_1 in the voltage divider, just as

before. What is the value of R_1? _____

– – – – – – – – – – – – – – – – –

R_1 will now be 280 ohms.

28. The value of C_1 is found the same as previously. What value of C_1 do

we get? _____

– – – – – – – – – – – – – – – – –

Using a time constant of 83.3 ms and a discharge resistance of 380 ohms, C_1 becomes 220 μF.

29. The voltage levels at the load resistor will now vary between 10.52 V
     and 9.47 V with a DC level of 10 V. These levels were found by using
     the voltage divider to find the resulting levels at the load for the 40 V_p
     value and also for V_x, which was 36 V for the full-wave rectifier. This
     is without a second smoothing capacitor.

     (a) If a second smoothing capacitor is used, what should its

          reactance be? ————————————————————————

     (b) What should its value be? (Remember that the waveform that it
          sees is now 120 Hz rather than 60 Hz as it was for half-wave recti-

          fication.) ————————————————————————

     ————————————————————

     (a) It should be one tenth (or less) of the load resistance, which
          means it should be 10 ohms or less.
     (b) Using 10 ohms and a frequency of 120 Hz, $C_2 = 135$ μF.

30. The AC voltage at the first smoothing capacitor will vary from 40 V to
     36 V. The AC output variation at the load will be as shown in frame 29
     without the second capacitor. With the addition of the second capacitor
     calculated in frame 29, what will be the AC variation at the output?

     ————————————————————————————————

     ————————————————————————————————

     ————————————————————

     $$\text{AC } V_{out} = (\text{AC } V_{in}) \times \frac{X_{C2}}{\sqrt{(R_1{}^2 + X_{C2}{}^2)}} = 0.143 \text{ V}$$

     This is about 0.14 V_{pp} which means that the output will now vary from
     10.07 to 9.93 V. Thus, the second capacitor has lowered the ripple even
     further. We can also see that the AC ripple is less than half of the
     amount of ripple that we saw for the half-wave rectifier shown in frame
     25. In other words, the full-wave rectifier produces an even smoother
     DC output.

31. Here is an example for you to work through. The circuit and input
     waveform are shown on the next page, and so is the value of the load and
     the required voltage across it. Find the values of the other components.
     Follow the steps laid out here as a guide.

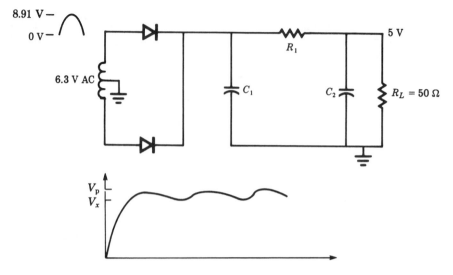

(a) What are V_p, V_x and the DC level at the first smoothing capacitor? _____

_____

(b) Calculate the value of R_1 to make the DC level at the output the required 5 V. _____

_____

(c) Now find the value of the first smoothing capacitor.

_____

(d) Now find the value of the second smoothing capacitor.

_____

(e) What is the final variation in the AC ripple at the output?

_____

(f) Now draw the final circuit showing the calculated values.

- - - - - - - - - - - - - - - - - - -

(a) $V_p = 6.3 \times \sqrt{2} = 8.91$ V, $V_x = 90\%$ of $V_p = 8.02$ V
    The DC level is 95% of V_p, which is 8.46 V.
(b) about 35 ohms
(c) 980 μF
(d) Using $X_{C_2} = 5$ ohms and 120 Hz, $C_2 = 265 \, \mu$F.
(e) At the input to the smoothing section, the AC variation is 8.91 to 8.02, or 0.89 V_{pp}. Using the AC voltage divider equation with $R_1 = 35$ ohms and $X_{C_2} = 5$ ohms, AC V_{out} equals about 0.13 V_{pp}. Therefore, the AC variation at the output will be 5.065 to 4.935 V. This is a very small AC ripple.

(f)

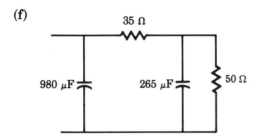

Using the simple procedure shown here will always produce a working power supply circuit. It is not the only design procedure for power supplies, but it is one of the simplest and most effective.

In summarizing this chapter a few disclaimers are in order.

All transformers are imperfect, so V_p shown at the secondary is not accurate.

Most common resistors have a 10% or sometimes a 20% tolerance in their stated value.

Electrolytic capacitors can be 50% low to 200% over in their stated value.

Because of the above factors, the time constant, voltage divider, and thus the residual ripple can never be exactly calculated.

In the light of the above, calculations made with textbook precision will never precisely agree with what is found in practice. Other approaches, which are much more rigorous in the mathematical sense, can be taken, but the outcome in practice is the same. What has been shown here will always work and will always produce an output within 10% of that desired.

For absolute accuracy of output voltage, a regulated power supply must be used. These are not covered in this book.

SELF-TEST

The questions below will test your understanding of this chapter. Use a separate sheet of paper for your diagrams or calculations. Compare your answers with the answers provided following the test.

In questions 1 through 5, draw the output waveshape of each circuit. The input is given in each case.

1.

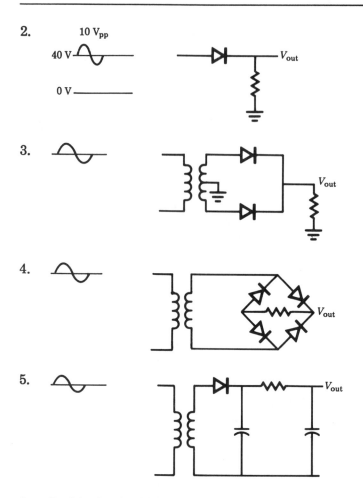

2.

3.

4.

5.

6. In this circuit, 100 V_{rms} at 60 Hz appears at the secondary of the transformer; 28 V DC with as little AC ripple as possible is required across the 220 ohm load. Find R_1, C_1, and C_2. Find the approximate AC ripple.

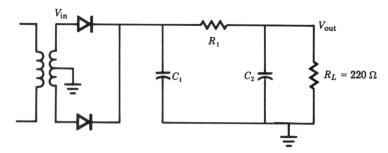

Answers to Self-Test

If your answers do not agree with those given below, review the frames indicated in parentheses.

1. 10 V
    0 V

    (frames 1–5)

2. 45 V
    40 V
    35 V

    (frame 2)

3.

    (frame 11)

4.

    (frame 13)

5.

    (frame 15)

6. $R_1 = 833$ ohms, $C_1 = 79$ μF,
    let $X_{C2} = 22$ ohms and then $C_2 = 60$ μF

    $$\text{AC } V_{\text{out}} = 14 \times \frac{22}{\sqrt{(22^2 + 833^2)}} = 0.37 \text{ V}_{\text{pp}}$$

    (frames 26–30)

Conclusion

What you have learned in this book are basics that will help you to continue studies in modern electronics. Modern electronics can be divided into two main branches: analog and digital. This book has laid the groundwork for each.

Transistor switching is the very simple principle on which the entire field of digital electronics lies. Digital circuits are merely several very simple transistor switching circuits used many times over. A computer may be able to do a lot of things and do them very fast, but it is merely a large number of simple transistor switching circuits.

The transistor amplifier is the basis of all analog, or linear, circuits. In this book we showed one important use in the amplifier and another in the oscillator. Although most analog circuits are now based on the IC op amp, (operational amplifier), this is simply made up of several transistor amplifiers.

The other chapters presented the basic behavior of the most important "passive" or non-amplifying components, and laid the foundation for further study.

You should now have enough knowledge to read many other elementary books, to follow what they are talking about, and to continue into electronics as far as you like and for whatever reason you wish. You should now be able to:

- recognize all the important electronics components;

- recognize several simple circuits;

- do simple circuit calculations;

- analyze simple circuits;

- design simple circuits;

- build simple circuits in a lab and make them work.

In summary what you have learned from this book is where electronics starts, that it follows two main paths, and that it is not very hard. To see how much you have learned, you may want to take the Final Self-Test that follows.

Final Self-Test

This final test will allow you to assess your overall learning of *Electronics.* Answers and review references are given following the test.

1. If $R = 1$ MΩ and $I = 2$ μA, find the voltage. _____

2. If $V = 5$ V and $R = 10$ kΩ, find the current. _____

3. If $V = 28$ V and $I = 4$ A, find the resistance. _____

4. If 330 ohms and 220 ohms are connected in parallel, find the equivalent resistance. _____

5. If $V = 28$ V and $I = 5$ mA, find the power. _____

6. If the current through a 220 ohm resistor is 30.2 mA, what is the power dissipated by the resistor? _____

7. If the power rating of a 1000 ohm resistor is 0.5 watts, what will be the maximum current that it can have? _____

8. If a 10 ohm resistor is in series with a 32 ohm resistor and the combination is across a 12 volt supply, what is the voltage drop across each resistor and what will the two drops add up to? _____

9. A current of 1 ampere splits between 6 ohm and 12 ohm resistors in parallel. Find the current through each. _____

10. A current of 273 mA splits between 330 ohms and 660 ohms as above. Find the current through each resistor. _____

11. If $R = 10$ kΩ and $C = 1$ μF, find the time constant. _____

12. If $R = 1$ MΩ and $C = 250$ μF, find the time constant. _____

13. Three capacitors, of 1 μF, 2 μF, and 3 μF are connected in parallel. Find the total capacitance. _____

14. Three capacitors, of 100 μF, 220 μF, and 220 μF, are connected in series. Find the total capacitance. _____

15. Three capacitors, of 22 pF, 22 pF, and 33 pF, are connected in series. Find the total capacitance. _____

16. What is the knee voltage for a germanium diode? _____

17. What is the knee voltage for a silicon diode? _____

18. In this circuit, $V_S = 5$ V, $R = 1$ kΩ. Find the current through the diode,

I_D. _____

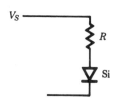

19. With the circuit in question 15, $V = 12$ V and $R = 100$ ohms. Find I_D.

_____

20. With this circuit $V_S = 100$ V, $R_1 = 7.2$ kΩ, $R_2 = 4$ kΩ, and $V_Z = 28$ V. Find the current through the zener diode, I_Z.

_____

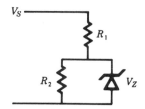

21. With the circuit in question 17, $V_S = 10$ V, $R_1 = 1$ kΩ, $R_2 = 10$ kΩ, and $V_Z = 6.3$ V. Find I_Z. _____

Use the circuit below for questions 22-24.

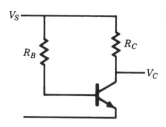

22. $V_S = 28$ V, $\beta = 10$, $R_B = 200$ kΩ, $R_C = 10$ kΩ. Find the DC collector voltage, V_C. _____

23. $V_S = 12$ V, $\beta = 250$, $R_C = 2.2$ kΩ, $V_C = 6$ V. Find R_B. _____
_____

24. $V_S = 10$ V, $R_B = 100$ kΩ, $R_C = 1$ kΩ, $V_C = 5$ V. Find β.
_____

25. What are the three terminals for a JFET called and which one controls the operation of the JFET? _____
When is the drain current at its maximum value? _____

Use this circuit for questions 26 and 27.

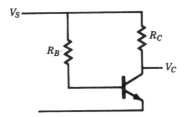

26. $V_S = 14$ V, $R_C = 10$ kΩ, $\beta = 50$. Find R_B to turn the transistor ON.
_____

27. $V_S = 5$ V, $R_C = 4.7$ kΩ, $\beta = 100$. Find R_B which will turn the transistor ON. _____

Use this circuit for questions 28 and 29.

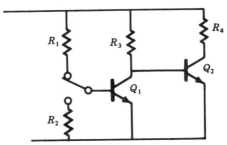

Find the values of R_1, R_2, and R_3 which will turn Q_2 ON and OFF.

28. $V_s = 10$ V, $\beta_1 = 50$, $\beta_2 = 20$, $R_4 = 2.2$ kΩ _____
_____

29. $V_s = 28$ V, $\beta_1 = 30$, $\beta_2 = 10$, $R_4 = 220$ Ω _____

30. An N-channel JFET has a drain saturation current of $I_{DSS} = 14$ mA. If a 28 V drain supply is used, what should the drain resistance, R_D, be when the JFET is in the ON state? _____

31. Draw one cycle of a sine wave.

32. Mark in V_{pp}, V_{rms}, and the period on your drawing for question 31.

33. If $V_{pp} = 10$ V, find V_{rms}. _____

34. If $V_{rms} = 120$ V, find V_{pp}. _____

35. If the frequency of a sine wave is 14.5 kHz, what is the period of the waveform? _____

36. Find the reactance X_C for a 200 μF capacitor when the frequency is 60 Hz. _____

37. Find the value of the capacitance which gives a 50 ohm reactance at a frequency of 10 kHz. _____

38. Find the inductive reactance X_L for a 10 mH inductor when the frequency is 440 Hz. _____

39. Find the value of the inductance which has 100 ohms reactance when the frequency is 1 kHz. _____

40. Find the series and parallel resonant frequency of a 0.1 μF capacitor and a 4 mH inductor that has negligible internal resistance. _____

For questions 41 and 42, use this circuit.

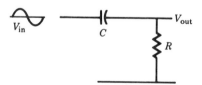

Find X_C, Z, V_{out}, I, $\tan \theta$ and θ.

41. $V_{in} = 10$ V_{pp}, $f = 1$ kHz, $C = 0.1$ μF, $R = 1600$ ohms.

42. $V_{in} = 120$ V_{rms}, $f = 60$ Hz, $C = 0.33$ μF, $R = 6$ kΩ.

For questions 43 and 44, use this circuit.

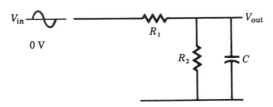

Find X_C, AC V_{out}, and DC V_{out}.

43. $V_{in} = 1$ V_{pp} AC, riding on a 5 V DC level; $f = 10$ kHz; $R_1 = 10$ kΩ;
$R_2 = 10$ kΩ; $C = 0.2$ μF.

44. $V_{in} = 0.5$ V_{pp} AC, riding on a 10 V DC level; $f = 120$ Hz; $R_1 = 80$ ohms;
$R_2 = 20$ ohms; $C = 1000$ μF.

45.   Use this circuit.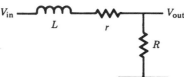

$V_{in} = 10$ V_{pp} AC, riding on a 5 V DC level; $f = 1$ kHz; $L = 10$ mH;
$r = 9$ ohms; $R = 54$ ohms.

Find AC V_{out}, DC V_{out}, X_L, Z, $\tan \theta$, and θ.

46.   In this circuit, $L = 1$ mH, $C = 0.1$ μF, and $R = 10$ ohms.

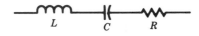

Find f_r, X_L, X_C, Z, Q, and the bandwidth.

47. In this circuit, $L = 10$ mH, $C = 0.02$ μF, and $r = 7$ ohms.

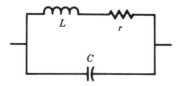

Find f_r, X_L, X_C, Z, Q, and the bandwidth.

48. If the voltage across the resonant circuit of problem 47 is at a peak value of 8 V at the resonant frequency, what is the voltage at the half power points? What are the half power frequencies? _____

Use this circuit for questions 49–51.

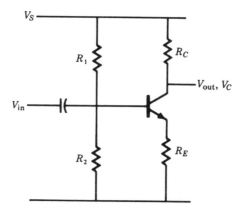

49. Design an amplifier to have a voltage gain of 10. Use $V_S = 28$ V, $R_C = 1$ kΩ, $\beta = 100$. _____

50. Design an amplifier to have a voltage gain of 20. Use $V_S = 10$ V, $R_C = 2.2$ kΩ, $\beta = 50$. _____

51. How would you modify the amplifier in question 50 to obtain a maximum gain? Assume the lowest frequency it has to pass is 50 Hz.

52. A JFET amplifier circuit is designed using the circuit shown in frame 42 of Chapter 8. If a bias point of $V_{GS} = -2.8$ V and a drain current of

$I_D = 2.7$ mA is selected, find the values of R_S and R_D. Choose a value of $V_{DS} = 12$ V.

_____

_____

53. If the transconductance of the JFET used in problem 52 is 4000 μmhos, what will the AC voltage gain be? _____

54. A certain op-amp circuit is using an input resistance of 8 kΩ to an inverting input. If the gain of the op-amp circuit is to be 85, what value should the feedback resistance be? _____

55. If the input to the op-amp circuit of problem 54 is 2 mV, what is the output? _____

56. What is an oscillator? _____

57. Why is positive feedback necessary rather than negative feedback in an oscillator? _____

_____

58. What feedback method is used in a Colpitts oscillator? _____

_____

59. What feedback method is used in a Hartley oscillator? _____

_____

60. Draw the circuit of a Colpitts oscillator.

61. Draw the circuit of a Hartley oscillator.

62. What is the frequency of the output of an oscillator? Give the formula.

_____

63. Draw a transformer circuit symbol. Include the center tap.

64. Name the two main turns used on a transformer. _____

65. What is the ratio of the voltage in to the voltage out? _____

66. How is the turns ratio related to the primary and secondary currents of the transformer? _____

67.   What is the impedance ratio between the turns? _____

68.   What are the two main uses for transformers? _____

_____   _____

69.   Draw a simple half-wave rectifier circuit with a smoothing filter at the output.

70.   Draw a simple full-wave rectifier circuit using a center tap transformer and a smoothing filter at the output.

71.   Design a full-wave rectified power supply with a 10 V transformer to give 5 V across a 50 ohm load. Use this circuit.

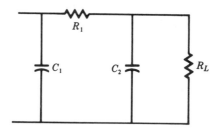

ANSWERS TO FINAL SELF-TEST

The references in parentheses following the answers give you the chapter and frame number where the material is introduced so you can easily review any material on the test.

1. $V = 2$ V   (Chapter 1, frame 5)

2. $I = 0.5$ mA   (Chapter 1, frame 6)

3. $R = 7$ ohms   (Chapter 1, frame 7)

4.   132 ohms   (Chapter 1, frame 10)

5. $P = 140$ milliwatts or 0.14 watts (Chapter 1, frames 13 and 14)

6.   0.2 W (Chapter 1, frames 13 and 15)

7.   22.36 mA (Chapter 1, frames 13 and 16)

8.   2.86 V, 9.14 V, 12 V (Chapter 1, frames 23 and 26)

9.   2/3 A through the 6 ohms; 1/3 A through the 12 ohms (Chapter 1, frame 27 or 29)

10.   91 mA through the 660 ohms; 182 mA through the 330 ohms (Chapter 1, frame 27 or 29)

11. $T = 0.01$ seconds (Chapter 1, frame 34)

12. $T = 250$ seconds (Chapter 1, frame 34)

13. $6 \ \mu F$ (Chapter 1, frame 40)

14. $52.4 \ \mu F$ (Chapter 1, frame 41)

15. $8.25 \ \mu F$ (Chapter 1, frame 41)

16. Ge = about 0.3 V (Chapter 2, frame 10)

17. Si = about 0.7 V (Chapter 2, frame 10)

18. $I_D = 4.3$ mA (Chapter 2, frame 14)

19. $I_D = 120$ mA (Chapter 2, frame 14)

20. $I_Z = 3$ mA (Chapter 2, frame 31)

21. $I_Z = 3.07$ mA (Chapter 2, frame 31)

22. $V_C = 14$ V (Chapter 3, frames 21–24)

23. $R_B = 1.1 \ M\Omega$ (Chapter 3, frames 21–24)

24. $\beta = 50$ (Chapter 3, frames 21–24)

25. drain, source, and gate, with the gate acting to control the JFET (Chapter 3, frame 29)

26. $R_B = 500 \ k\Omega$ (Chapter 4, frames 4–8)

27. $R_B = 470 \ k\Omega$ (Chapter 4, frames 4–9)

28. $R_3 = 44 \ k\Omega$, $R_1 = 2.2 \ M\Omega$, $R_2 = 1 \ k\Omega$ to $1 \ M\Omega$ ($1 \ M\Omega$ is the best choice) (Chapter 4, frames 19–23)

29. $R_3 = 2.2 \ k\Omega$, $R_1 = 66 \ k\Omega$, $R_2 = 66 \ k\Omega$ (Chapter 4, frames 19–23)

30. $R_D = 2 \ k\Omega$ (Chapter 4, frame 39)

31.

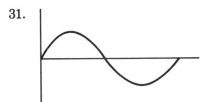

(Chapter 5, frame 3)

32.

(Chapter 5, frames 3 and 7)

33. 3.535 V (Chapter 5, frame 4)

34. 340 V (Chapter 5, frame 5)

35. 69 μsec (Chapter 5, frame 7)

36. 13.3 ohms (Chapter 5, frame 14)

37. 0.32 μF (Chapter 5, frame 14)

38. 27.6 ohms (Chapter 5, frame 17)

39. 16 mH (Chapter 5, frame 17)

40. 8 kHz (Chapter 5, frames 19 and 20)

41. $X_C = 1.6$ kΩ, $Z = 2263$ ohms, $V_{out} = 7.07$ V, $I = 4.4$ mA, tan $\theta = 1$, $\theta = 45$ degrees (Chapter 6, frames 10 and 23)

42. $X_C = 8$ kΩ, $Z = 10$ kΩ, $V_{out} = 72$ V, $I = 12$ mA, tan $\theta = 1.33$, $\theta = 53.13$ degrees (Chapter 6, frames 10 and 23)

43. $X_C = 80$ ohms, AC $V_{out} = 8$ mV, DC $V_{out} = 2.5$ V (Chapter 6, frame 26)

44. $X_C = 1.33$ ohms, AC $V_{out} = 8.3$ mV, DC $V_{out} = 2$V (Chapter 6, frame 26)

45. $X_L = 62.8$ ohms, $Z = 89$ ohms, AC $V_{out} = 6.07$ V, DC $V_{out} = 4.3$ V, tan $\theta = 1$ (approximately), $\theta = 45$ degrees (Chapter 6, frames 31 and 35)

46. $f_r = 16$ kHz, $X_L = 100$ ohms, $X_C = 100$ ohms, $Z = 10$ ohms, $Q = 10$, BW = 1.6 kΩ (Chapter 7, frames 2, 6, and 20)

47. $f_r = 11{,}254$ Hz, $X_L = X_C = 707$ ohms, $Z = 71.4$ kΩ, $Q = 101$, BW = 111 Hz (Chapter 7, frames 10, 11, and 20)

48. $V_{hp} = 5.656$ V, $f_{1hp} = 11{,}198$ Hz, $f_{2hp} = 11{,}310$ Hz (Chapter 7, frame 27)

49. These values will work. Your values should be close to these. $R_E = 100$ ohms, $V_C = 14$ V, $V_E = 1.4$ V, $V_B = 2.1$ V, $R_2 = 1.5$ kΩ, $R_1 = 16.8$ kΩ (Chapter 8, frame 17)

50. $R_E = 110$ ohms, $V_C = 5$ V, $V_E = 0.25$ V, $V_B = 0.95$ V, $R_2 = 2.2$ kΩ, $R_1 = 18.1$ kΩ (Chapter 8, frame 17)

51. The gain can be increased by using a capacitor to bypass the emitter resistor R_E; $C_E = 300$ μF (approximately) (Chapter 8, frame 20)

52. $R_S = 1.04$ kΩ, $R_D = 3.41$ kΩ (Chapter 8, frame 42)

53. $A_v = -13.6$ (Chapter 8, frame 39)

54. $R_F = 680$ kΩ (Chapter 8, frame 45)

55. $V_{out} = 170$ mV and is inverted (Chapter 8, frame 45)

56. An oscillator is a circuit which emits a continuous sine wave output, without the need for an input signal. Other types of oscillators exist

which do not have sine wave outputs, but we did not cover them in this book.   (Chapter 9, introduction)

57. Positive feedback will cause the amplifier to sustain an oscillation or sine wave at the output. Negative feedback will cause the amplifier to stabilize, which reduces oscillations at the output. (Chapter 9, frames 2 and 3)

58. a capacitive voltage divider   (Chapter 9, frame 14)

59. an inductive voltage divider   (Chapter 9, frame 14)

60.

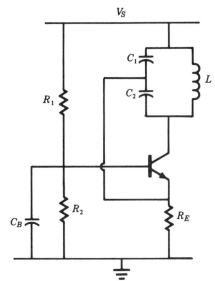

(Chapter 9, frame 24)

61.

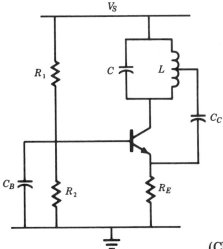

(Chapter 9, frame 25)

62. $f_r = \dfrac{1}{2\pi\sqrt{LC}}$ (Chapter 9, frame 11)

63.

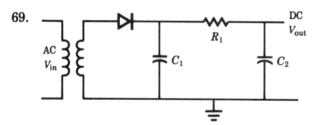

Pri Sec

(Chapter 10, frame 4)

64. primary and secondary (Chapter 10, frame 2)

65. $V_{in}/V_{out} = V_P/V_S = N_P/N_S = TR$ (Chapter 10, frame 6)

66. $I_{out}/I_{in} = I_S/I_P = N_P/N_S = TR$ (Chapter 10, frame 16)

67. $Z_{in}/Z_{out} = (N_P/N_S)^2$, or impedance ratio is the square of the turns ratio (Chapter 10, frame 16)

68. stepping up or stepping down an AC voltage, and for connecting communications circuits (Chapter 10, introduction)

69.

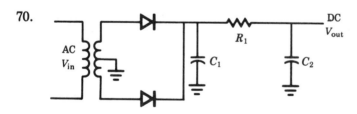

(Chapter 11, frame 14)

70.

(Chapter 11, frame 31)

71. $R_1 = 84$ ohms, $C_1 = 622\ \mu F$, $C_2 = 265\ \mu F$ (Chapter 11, frames 26–29)

Appendix

This appendix includes a list of terms used in the book as well as a list of symbols and abbreviations. In addition, there are two tables, one showing the powers of ten with their standard engineering metric prefixes and the other listing standard composition resistor values.

The definitions given for the terms listed are intended to convey a general idea of the meaning and sense in which they are used here. Although more terms and definitions might be found in some textbooks, the ones shown here are the most commonly used at this level of electronics. The symbols and abbreviations employed are generally accepted in the field of electronics for identifying the item in question. The tables contain standard information and are for reference. Finally, there is a supplemental reading list containing books that may be of interest to the reader.

I. Terms

Ampere (A): The unit of current.

Amplifier: Electronic device or circuit that produces at the output a larger power, voltage, or current for a smaller power, voltage, or current at the input. In this book we see amplifiers used to produce an amplified version of an input waveform.

Capacitance (C): The ratio between the electric charge transferred from one electrode of a capacitor to another and the resultant difference in potential between the electrodes.

Capacitor: A component that stores electric charge which can then be returned to a circuit in the form of electric current.

Diode: A component that will conduct current in one direction only. The modern diode is two semiconductor pieces joined together to form a junction.

Farad (F): Unit of capacitance. 1 farad exists when a voltage variation of 1 volt per second induces 1 ampere.

Feedback: In electronics, a connection from the output of an amplifier back to the input, whereby a portion of the output voltage is used to control, stabilize, or modify the operation of the amplifier. In this book we see feedback used in oscillators and the op-amp circuit.

Frequency (f): Number of recurrences of a periodic signal in a given time period, measured in hertz.

Ground: Zero volts. The arbitrary reference point in a circuit from which all voltage measurements are made. Usually the negative terminal of the battery is connected to this point, and in many cases this point is also physically connected to the earth or ground via a water pipe.

Henry (H): Unit of inductance. 1 henry inductance exists when a current variation of 1 ampere per second induces 1 volt.

Impedance (Z): Total opposition (resistance and reactance) by a circuit to AC current flow, measured in ohms.

Inductance (L): Property of a circuit which opposes change in an existing current.

Inductor: A coil of wire, whose magnetic field opposes changes in current flow when the voltage applied is changed.

Kirchhoff's laws: Form the basis for DC and AC circuit analysis. *Current law (KCL):* The summation of all currents at a junction equals zero. *Voltage law (KVL):* The summation of all voltages in a loop equals zero.

Ohm's law: A formulation of the relationship of voltage, current, and resistance, expressed as $E = IR$.

Operational amplifier (op-amp): Electronic amplifier circuit, generally packaged as an integrated circuit, that has a multitude of applications. It has high input impedance with inverting and noninverting input terminals, low output impedance, and exhibits high gain.

Oscillator: Electronic circuit built around an amplifier with positive feedback that produces a sine wave at the output.

Phase angle: In an AC circuit, the angle of lead or lag between the current and voltage waveforms. It is also the angle of the impedance of the circuit.

Phase shift: Refers to the phase change of a signal as it passes through a circuit such as an amplifier.

Power: The expenditure of energy over time. Units are watts, which are further defined as energy (joules) per second.

Reactance (X): Opposition to flow of alternating current, measured in ohms. Capacitive reactance (X_C) and inductive reactance (X_L) are offered by capacitors and inductors, respectively.

Rectification: Process of producing a DC output from an AC input waveform.

Resistance (R): The quality possessed by conductors of electricity which causes them to oppose the flow of current when a battery is connected.

Semiconductor: A material whose electrical characteristics fall between those of a conductor and those of an insulator. The two most common materials are silicon and germanium.

Transformer: Highly efficient magnetic circuit device that transforms an input AC voltage to either a higher level (step up) or a lower level (step down) AC voltage. The transformer input power is generally considered equal to the transformer output power.

Transistor, BJT: The bipolar junction transistor (BJT) is made of two semi-conductor PN junctions joined together with special construction to give certain characteristic properties. The final component is a three terminal device (emitter, base, and collector) and is either in an NPN or PNP configuration. In this book we use mainly the NPN, solely for the sake of simplicity. Generically, the word transistor is used for the BJT whenever one is referring to it.

Transistor, JFET: The junction field effect transistor (JFET) is made with semiconductor materials in a construction that yields either an N-channel or a P-channel as the primary current path through the device. The final component is a three terminal device (drain, gate, and source). In this book we have used mainly the N-channel JFET in our examples and discussions.

Turns ratio (TR): The ratio of the number of turns in the primary or input winding of a transformer to the number of turns in the secondary or output winding.

Volt (V): This term has been used extensively but has not been formally defined. As far as we are concerned it is the "force" which causes a current to flow through a conductor.

Watt (W): Unit of electric power required to do work at the rate of 1 joule per second.

Zener: A particular type of diode which "breaks down" at a definite voltage level.

II. Symbols and Abbreviations

A: Ampere

AC: Alternating current.

A_V: Voltage gain.

β (beta): The ratio of I_C/I_B, called the current gain.

BW: The bandwidth, which is the range of frequencies which will be passed or rejected by an LC circuit.

DC: Direct current.

E: Voltage symbol used in Ohm's law.

F: Farad.

f_r: The resonant frequency of an LC circuit.

H: Henry.

Hz: Hertz.

I: Electron current through a conductor. Measured in amperes.

I_B: The base current of a transistor.

I_C: The collector current of a transistor. In this book both the collector current and the base current have been presented as DC quantities which are part of the biasing process.

I_D: The drain current of a JFET.

I_{DSS}: The drain current of a JFET when it is at full conduction or saturation.

LC: Inductor-capacitor circuit.

N_p, N_s: Number of turns in primary and secondary coils of a transformer.

Ω: Ohm symbol.

Ohm (Ω): The unit of measurement for resistance and reactance.

P: Power

Q: The measure of how "good" a circuit is at passing or rejecting a band of frequencies.

R: Resistance.

r: The DC resistance of an inductor.

T: Period of a waveform.

TC: Time constant.

V: Voltage.

V_C: The DC voltage level at the collector of a transistor.

V_{DD}: The drain supply voltage in a JFET circuit.

V_E: The DC voltage at the emitter of a transistor.

V_{GS}: The gate to source voltage for a JFET.

$V_{GS(off)}$: The cutoff voltage for a JFET where the drain current is zero.

V_{in}: The AC signal input voltage to a circuit. Often this is centered around 0 volts, but sometimes it is centered around or "riding on" a DC level above ground.

V_{out}: The AC output voltage of a circuit. Sometimes this has been at the collector of a transistor, but not always.

V_{pp}: The voltage range from the negative to the positive peaks of a signal. In this book sine waves only have been used, but this term can be used with all waveforms.

V_{rms}: This applies to sine waves only. Defined as $V_{rms} = 0.707 \, V_{pp}/2$.

V_S: The supply voltage to a circuit.

W: Watt.

X_C: The reactance of a capacitor. This depends on frequency.

X_L: The reactance of the inductor. This depends on frequency.

Z: Impedance.

III. Powers of Ten and Engineering Prefixes

Power	Decimal	Prefix	Symbol	Example
10^9	1,000,000,000	Giga-	G	Gohms
10^6	1,000,000	Mega-	M	Mohms
10^3	1,000	Kilo-	k	kohms
10^{-3}	0.001	Milli-	m	mamp
10^{-6}	0.000,001	Micro-	μ	μamp
10^{-9}	0.000,000,001	Nano-	n	nsec
10^{-12}	0.000,000,000,001	Pico-	p	pF

IV. Standard Composition Resistor Values

The values shown here are commonly considered to be the standard values. However, suppliers do not often stock all values or have values in all tolerances or power ratings. Composition resistors are generally of the carbon type and are the most common resistor.

Ohms	Ohms	Ohms	kohms	kohms	Mohms
1.0	27	680	10	270	1.0
1.1	30	750	11	300	1.1
1.2	33	820	12	330	1.2
1.3	36	910	13	360	1.3
1.5	39	1000	15	390	1.5
1.6	43	1100	16	430	1.6
1.8	47	1200	18	470	1.8
2.0	51	1300	20	510	2.0
2.2	56	1500	22	560	2.2
2.4	62	1600	24	620	2.4
2.7	68	1800	27	680	2.7
3.0	75	2000	30	750	3.0
3.3	82	2200	33	820	3.6
3.6	91	2400	36	910	3.9
3.9	100	2700	39		4.3
4.3	110	3000	43		4.7
4.7	120	3300	47		5.1
5.1	130	3600	51		5.6
5.6	150	3900	56		6.2
6.2	160	4300	62		6.8
6.8	180	4700	68		7.5
7.5	200	5100	75		8.2
8.2	220	5600	82		9.1
9.1	240	6200	91		10.0
10	270	6800	100		11.0
11	300	7500	110		12.0
12	330	8200	120		13.0
13	360	9100	130		15.0
15	390		150		16.0
16	430		160		18.0
18	470		180		20.0
20	510		200		22.0
22	560		220		
24	620		240		

V. Supplemental Reading

The following list of books may be of interest to the reader who is seeking more knowledge or information regarding basic electricity, circuit analysis, or electronic devices and design. The list contains texts that present material in a simple fashion as well as a few that are a little more rigorous. After reading and understanding this text, you should be ready for any of the books listed here. The list is by no means exhaustive; many other texts could be found in any library.

Bell. *Electronic Devices and Circuits.* 2nd ed. Reston Publishing Company, Inc., 1980.

Boylestad. *Introductory Circuit Analysis.* 4th ed. Charles E. Merrill Publishing Company, 1982.

Faber. *Essentials of Solid State Electronics.* John Wiley & Sons, Inc., 1985.

Faulkenberry. *An Introduction to Operational Amplifiers.* Prentice-Hall, Inc., 1977.

Hughes. *Basic Electronics, Theory and Experimentation.* Prentice-Hall, Inc., 1984.

Johnson and Jayakumar. *Operational Amplifier Circuits, Design and Application.* Prentice-Hall, Inc., 1982.

Oppenheimer. *Fundamentals of Electric Circuits.* Prentice-Hall, Inc., 1984.

Patrick and Fardo. *Electricity and Electronics.* Prentice-Hall, Inc., 1984.

Rutkowski. *Basic Electricity for Electronics, A Text-Laboratory Manual.* Bobbs-Merrill Educational Publishing, 1984.

Ryan. *Basic Electricity: A Self-Teaching Guide.* John Wiley & Sons, Inc., 1976.

Ryder and Thomson. *Electronic Circuits and Systems.* Prentice-Hall, Inc., 1976.

Schwartz. *Survey of Electronics.* 3rd ed. Charles E. Merrill Publishing Company, 1985.

Taber and Silgalis. *Electric Circuit Analysis.* Houghton Mifflin Company, 1980.

Williams. *Electronics for Everyone.* Science Research Associates, Inc., 1979.

Zeines. *Transistor Circuit Analysis and Application.* Reston Publishing Company, Inc., 1976.

Index